Control
of
Cardiac Rhythm

Control
of
Cardiac Rhythm

E. M. Vaughan Williams, D.M., D.Sc., F.R.C.P.

Emeritus Fellow
Hertford College
Oxford University
Oxford, United Kingdom

and

John C. Somberg, M.D.

Professor of Medicine and Pharmacology
Finch University of Health Sciences
The Chicago Medical School
North Chicago, Illinois

Lippincott - Raven
P U B L I S H E R S
Philadelphia • New York

Midterm

Acquisitions Editor: Ruth W. Weinberg
Developmental Editor: Judith E. Hummel
Manufacturing Manager: Dennis Teston
Production Manager: Lawrence Bernstein
Production Editor: Lawrence Bernstein
Cover Designer: Kevin Kall
Indexer: Phyllis Manner
Compositor: Lippincott–Raven Electronic Production
Printer: Maple Press

Printed in the United States of America

9 8 7 6 5 4 3 2 1

Library of Congress Cataloging-in-Publication Data

Vaughan Williams, E. M.
   Control of cardiac rhythm / by E.M. Vaughan Williams and John C. Somberg.
      p.   cm.
   Includes bibliographical references and index.
   ISBN 0-397-58783-X
   1. Myocardial depressants.   I. Somberg, John C.   II. Title.
   [DNLM: 1. Arrhythmia—prevention & control.   2. Arrhythmia—drug
therapy.   3. AntiArrhythmia Agents—pharmacology.   WG 330 V371c 1997]
   RM347.V383 1997
616.1'28061--dc21
DNLM/DLC
for Library of Congress

2/28/00 H.S.

Contents

10. The Clinical Pharmacology of Antiarrythmic Agents **115**
John C. Somberg

Preface

The aim of this book is to provide a thorough historical review and an up-to-date account of the development and current usage of antiarrhythmic therapy. It includes more than 770 references, half of which are from 1995 and 1996. Our policy has been, wherever possible in referring to papers, to quote the actual words of the authors themselves.

The first chapter presents the historical background from the beginnings of cardiology to the present day. The final chapter assesses current practice, and looks to developments likely in the future.

Chapters 2, 3, 4, 6, and 7 are each devoted to an explanation of one of the five classes of antiarrhythmic action, and describe the electrophysiologyical effects and clinical use of the major drugs in each category. The background basic science (effects on ionic currents, molecular structure of ion channels, etc.) is given in each section, so that by the end of the book the reader will have been introduced to a full understanding of cardiac electrophysiology. The efficacy and side effects of individual compounds are also described. Thus this book will enable physicians to make rational choices of therapy to suit their diagnoses.

The remaining chapters deal with the current most-discussed topics on the origin and treatment of cardiac arrhythmias: the aftermath of CAST, the genetic defects in the Long QT Syndrome, the arrhythmic risk in ischemia and hypertrophy, and in cardiac failure, torsade de pointes, proarrhythmia and how to avoid it, ischemic preconditioning, etc. A comprehensive discussion of the clinical pharmacology of currently available anti-arrhythmic agents in the United States is provided in Chapter 10, followed by a clinical therapeutic approach to cardiac arrythmias in chapter 11.

The book is written for several different groups: medical students with an interest in cardiology, cardiology fellows, and research scientists in industry working in the cardiovascular field. Consultant cardiologists will also find the book useful for brushing up their acquaintance with recent developments in arrhythmology, as well as an up-to-date comprehensive approach to arrhythmia therapeutics.

Control
of
Cardiac Rhythm

1

Historical Background

E. M. Vaughan Williams

The antiarrhythmic action of the first drug able to control a cardiac arrhythmia was discovered by accident, as described in the following account quoted from an early review (1).

Although de Senac employed cinchona for the treatment of rebellious palpitation in the middle of the eighteenth century, the introduction of quinine as an antifibrillatory drug is usually attributed to Wenckebach (1914). Until this time atrial fibrillation had been regarded as a life sentence, and most clinicians had restricted themselves to reassuring their patients that although no effective treatment was known, the condition was compatible with many years of useful life. One of Wenckebach's patients, however, declined to be reassured, and declared that he was better informed than his physician. He promised to return the following day and vowed that his fibrillation would have stopped. He did, and it had. Wenckebach walked over to lock the door and placing the key in his pocket said, "You do not leave this room until you have told me how you did that." It turned out that the patient was a merchant whose business took him to parts of the world where malaria was endemic, and he was accustomed to take quinine. He had noticed that his fibrillation was sometimes arrested after a large dose. Subsequently Frey (1918) showed that the stereoisomer quinidine was even more effective.

The British founder of the journal *Heart*, Thomas Lewis, carried out extensive studies of the action of quinidine on dog hearts and concluded (1921) that the refractory period was prolonged. There was controversy in subsequent years about the mechanism, some claiming that the duration of the cardiac action potential was increased, others that it was not, but techniques capable of resolving the problem did not become available until 30 years later. After World War II the introduction of intracellular electrodes longitudinally inside squid giant axons had led to the discovery that the depolarization phase of the action potential was due to a transfer of positive charge by a net influx of sodium ions, followed, after a brief delay, by efflux of potassium ions (delayed rectification), which repolarized the axon. Such techniques were obviously inapplicable to cardiac muscle (Fig. 1).

Gilbert Ling, working in Gerard's laboratory, revolutionized electrophysiology by producing glass capillary microelectrodes with tips 0.1 to 0.5 microns in diameter, which could be inserted directly into the interior of cells (2). It was soon shown that sodium ions were responsible for depolarization in cardiac Purkinje cells. At that time UK investigators without the good fortune of access to US equipment who wished to use Ling's method faced the unavailability of stimulators and DC amplifiers and were obliged to construct their own apparatus with components (purchased by weight regardless of function or serviceability) salvaged from defunct tanks and aircraft. To study contracting cardiac muscle, in contrast to motionless Purkinje cells, it was also necessary to devise a flexible suspension system for the microelectrode so that the tip could remain inside a moving fiber.

The use of microelectrodes in spontaneously contracting rabbit atrial muscle revealed that quinidine, over the range of clinical concentrations, reduced the maximum velocity of depolarization (V_{max}), with only a trivial or absent effect on action potential duration (APD). Even ten times a clinical concentration had no effect on the resting

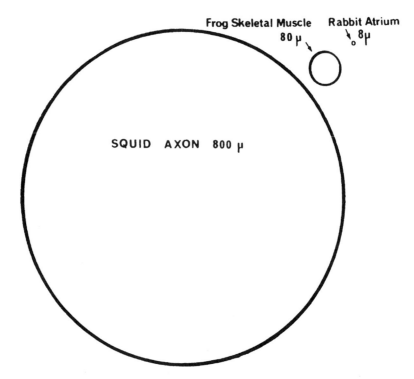

Frog Skeletal Muscle   Rabbit Atrium
80 μ   8μ

SQUID   AXON   800 μ

FIG 1. Diameters of squid axon, frog skelatal muscle and rabbit atrial muscle drawn to scale.

potential (3). The effect on depolarization was frequency dependent; the faster the frequency the greater the effect. Why, then, had some earlier investigators found that quinidine prolonged APD? An explanation was provided by West and Amory (4) who observed that the effect of quinidine in delaying repolarization was observed only at abnormally low frequencies; at normal or higher frequencies the effect was absent at clinical concentrations. Thus the effect of quinidine on APD exhibited reverse frequency dependence, a phenomenon recently rediscovered.

ACTION POTENTIAL PLATEAU

In nerves the APD (and refractory period) is very short, to enable impulses to be carried at high frequency. In skeletal muscle APD also is short, because contraction-coupling effects summate, inducing a tetanic contraction at the fusion frequency appropriate to the type of muscle (from 30 Hz in red muscle to 300 Hz in eye muscle). It had already been deduced from differential recordings with extracellular electrodes that the action potential of cardiac was of a prolonged unitary type originally described by Bozler (reviewed, 5) Microelectrode recordings from various cardiac tissues revealed a wide range of APDs; atrial < ventricular epicardial < ventricular endocardial < proximal Purkinje cell < distal Purkinje cell. The mechanism by which the action potential plateau was maintained had yet to be elucidated.

THE SECOND INWARD CURRENT

Quinidine restricted sodium entry and the rate of rise of the action potential became progressively slower, but before depolarization failed altogether the upstroke developed a step. Provided the initial phase of depolarization reached a level positive to approximately

−40 mV, depolarization continued more slowly to an overshoot (positive to zero).

"A fast phase of depolarization was cut off early, yet depolarization proceeded at a slower rate to an overshoot. The early fast phase of the step represents the operation of the sodium-carrier, and the second slower phase is produced by *current carried by other ions*. There is evidence that in some tissues ions other than sodium can enter the cell during depolarization" (3).

The evidence referred to in the above quotation was that of Fatt and Katz (6), who had demonstrated that in crustacean muscle fibers "the influx of calcium . . . may be responsible for the transport of charge."

Anomalous (Inward) Rectification

The cytoplasm of excitable tissues contains a high concentration of potassium (K) ions (about 145 mM), but low sodium (Na, about 10 mM), established by the ATP-dependent Na/K pump, which extrudes three Na ions in exchange for two K ions, creating one net negative charge per cycle. At rest the membrane is so much more permeable to K than to other ions that the resting potential is close to the K equilibrium potential E_K. A study of the relation between extracellular K concentration, $[K]_o$, and resting potential in rabbit atrial muscle, over a range of $[K]_o$ concentrations (1 to 11 mM) at which the atria were still able to beat spontaneously, revealed that the resting potential approximated more closely to E_K in high $[K]_o$ than in low $[K]_o$ (7). As $[K]_o$ decreased, E_K became progressively more negative to the recorded potential (E), suggesting that K permeability decreased when the intracellular potential was positive to E_K. This would clearly be of advantage to the cell, because when E is positive to E_K, potassium would flow out of the cell and would have to be pumped back in again. The effect is the opposite of delayed rectification, whereby K permeability is increased on depolarization, in order to repolarize the cell. A similar phenomenon was reported in the same year by

Hodgkin and Horowitz (8) in frog muscle and was initially termed anomalous rectification, in contrast to the outward repolarizing current of delayed rectification. Subsequently it has been referred to as inward rectification, reinforcing depolarization when the current cuts out as E goes positive to E_K.

In 1959, therefore, two phenomena had been described that could account for the maintenance of a depolarized plateau in cardiac muscle: a second slow inward current carried by calcium ions, and a reduction of potassium permeability. In 1962 Noble (9) showed that Purkinje cell action potentials could be simulated by modifications of the Hodgkin-Huxley equations derived to explain the squid axon action potential. The cationic charge carriers were assumed to be Na and K ions only, and although anomalous rectification was invoked to explain the plateau, no second inward or calcium current was included. Nevertheless a remarkably close fit was obtained between the calculated and observed potentials, which illustrates the futility of computations based on inadequate data.

CLASS 1 ANTIARRHYTHMIC ACTION

Antiarrhythmic effects of some drugs other than quinidine had been reported by 1960, and the technique of intracellular recording, in addition to other methods, was employed for a systematic study of their mode of action (10). Their electrophysiological effects were so similar that it seemed justifiable to conclude that the antiarrhythmic action was due to a property they all shared (although they might differ in other respects). None of the compounds altered the resting potential, even at high concentration.

"All the drugs studied had in common the ability to slow the rate of rise of the action potential at low concentrations, and it has been concluded that this is the property that confers their antiarrhythmic action. The drugs interfere specifically with the process by which depolarizing charge is transferred across the membrane." Consequently "they extended the refractory period to a point *long after the time at which repolarization was already complete*" (10).

Another consequence of this effect was to increase the electrical threshold at which cardiac muscle could respond to a stimulus.

Thus was defined the first (and at that time the only) class of antiarrhythmic action, interference with the transfer of depolarizing charge (carried by sodium ions). The effect was antiarrhythmic: (a) by prolonging refractory period, (b) by raising electrical threshold and so protecting *normal* myocardium from invasion by ectopic impulses, and (c) by causing conduction failure preferentially in abnormal (partly depolarized) muscle.

CLASS 2 ANTIARRHYTHMIC ACTION

For many years it had been recognized that abnormalities of cardiac rhythm could be precipitated or exacerbated by adrenergic stimuli and that the induction of ventricular tachycardia (VT) or fibrillation (VF) was a danger in the presence of adrenergic "sensitizing agents" such as halogen anesthetics. Antisympathetic drugs available in the nineteen-sixties, however, acted mainly on alpha adrenoceptors. The first beta adrenoceptor antagonist used clinically, pronethalol (nethalide), was introduced in 1962 (11) and was soon shown to have antiarrhythmic properties (12). It so happened, however, that pronethalol was also a local anesthetic (13), as was its successor, propranolol, and both compounds had a Class 1 action on cardiac tissues, which led some authors to doubt whether their antisympathetic effect could be responsible for the antiarrhythmic action. The question was settled by a comparison of (–) and (+) isomers of propranolol (14), which were equal in potency for their local anesthetic and Class 1 activities, but the (–) isomer had much greater efficacy both as a beta blocker and as an antiarrhythmic. Furthermore, the antisympathetic agent bretylium, which first releases then blocks the release of noradrenaline from sympathetic nerve endings, was shown to be antiarrhythmic, although almost devoid of local anesthetic or Class 1 activity (15). Thus a second class of antiarrhythmic action was established, antagonism to the arrhythmogenicity of adrenergic stimulation, the mechanism being either blockade of adrenoceptors or prevention of noradrenaline release.

CLASS 3 ANTIARRHYTHMIC ACTION

The antiarrhythmic and antihypertensive effects of beta blockers had not been anticipated at the time of their introduction for the treatment of ischemic heart disease (IHD). Initially there were fears that they might be contraindicated in patients with impaired cardiac contractility, and the Belgian firm Labaz produced amiodarone—an antisympathetic agent for the treatment of angina pectoris—which was *not* a beta blocker. The mechanism of its action was unclear, but was thought to be due to a post-receptor intracellular action on the pathway of effects mediated by both alpha and beta adrenoceptors. Amiodarone contains iodine that is split off, and there was a possibility of interference with thyroid function. Meanwhile it had been shown that the hyperthyroid state shortens cardiac APD, whereas thyroidectomy lengthens it (16). Amiodarone administered to rabbits long term greatly prolonged APD in atria and ventricles, but this was not due to any iodine released, nor was the drug a competitive antagonist of thyroid hormone (17). Delay of repolarization automatically prolongs refractory period, and amiodarone was shown to possess an antiarrhythmic action more potent than could be accounted for by a very minor Class 1 action. A third class of antiarrhythmic action was therefore defined as a uniform and homogeneous delay of repolarization.

Amiodarone prolonged the QT interval of the electrocardiogram (ECG). QT prolongation, however, was often associated with an increased probability of arrhythmia, and many clinicians questioned whether a drug that prolonged QT could be antiarrhythmic. The QT interval can be prolonged in different ways, one of which is a greater dispersion of repolarization times, reflecting heterogeneous myocardial damage. Heterogeneity of delayed conduction could also give rise to dispersion of repolarization times without any change in APD. Amiodarone, in contrast, produced a homogeneous prolongation of APD, which was antiarrhythmic.

These studies of the mode of action of antiarrhythmic drugs, accumulated over 13 years, led to the publication in 1970 of a classification of antiarrhythmic actions (18).

CLASS 4 ANTIARRHYTHMIC ACTION

The second slow depolarizing current in cardiac muscle, carried by ions other than sodium, probably calcium, had been identified in 1958 (3), but the evidence was ignored for a decade, presumably because the current involved had not been characterized by voltage-clamp experiments. The drug verapamil had been introduced as a coronary artery dilator for the treatment of angina, but was found empirically to have antiarrhythmic properties in some patients. The mode of action was unknown, but electrophysiological studies indicated that the antiarrhythmic effect could not be attributed to any of three classes of action already described, and it was suggested that verapamil exerted an entirely new class of action (19). It had a negligible Class 1 action, but caused an atropine-resistant slowing of sinoatrial node (SAN) frequency and depressed contractions in atria and ventricles and reduced the positive inotropic effect of raised extracellular calcium $[Ca]_o$. It did not lengthen APD and had no specific antisympathetic effect.

"Fleckenstein et al. (1968) suggested that verapamil might act by blocking 'the movements of calcium from the excited fiber membrane to the contractile system or by competing with Ca for active sites where ATP is split.' Their observations pertain to potential-contraction coupling which is an issue quite distinct from the possibility that calcium may carry depolarizing current. There is good evidence that calcium-carried spikes—that is, fast depolarizations—can occur in smooth muscle (Brading et al. 1969). Many years ago it was suggested that in cardiac muscle, in addition to the early sodium-carried depolarizing current, there might be a second later inward current carried by ions other than sodium. . . . Interference by a drug such as verapamil with inward calcium movement across the membrane would not only account for its negative inotropic action, but would also constitute a fourth class of anti-dysrhythmic action" (19).

An inward plateau current of calcium had meanwhile been demonstrated by voltage clamp methods, and it was subsequently confirmed that verapamil did indeed block this current, and the era of calcium antagonists began. Later studies revealed that depolarizing current in the SAN and atrioventricular node (AVN) was carried not by sodium but by calcium, which explained why verapamil had been found clinically to be especially effective in supraventricular arrhythmias.

CLASS 5 ANTIARRHYTHMIC ACTION

In 1979 it was reported that alinidine, the N-allyl derivative of the antihypertensive drug clonidine, exerted a specific bradycardic action on the sinus node and had an antiarrhythmic action in some experimental models of arrhythmia (20). Detailed electrophysiological studies revealed that alinidine reduced the slope of the slow diastolic depolarization in pacemaking cells of the SAN. The effect was unaltered by atropine. Alinidine did not block sodium channels or adrenoceptors, nor did it increase potassium conductance (as does acetylcholine [Ach]). There was no negative inotropic action, and the positive inotropic effect of raised $[Ca]_o$ was unchanged. Thus, since alinidine possessed none of the four antiarrhythmic actions already described, it was necessary to propose a fifth class of action (21–23). The composition of ionic currents contributing to SAN diastolic depolarization is complex and controversial, and it is discussed in detail later, but recent studies of alinidine and its successors, the specific bradycardic agents (SBA), have confirmed that their action differs fundamentally from that of other antiarrhythmic drugs, justifying the proposal that reduction of the slope of the slow diastolic depolarization in the SAN constitutes a distinct fifth class of action.

SUBSEQUENT DEVELOPMENTS

Since 1970 the expansion of knowledge about the universe in general and the biosphere in particular has been spectacular and continues at such a pace that it is surprising that no sixth class of antiarrhythmic action has emerged. In a visible Universe backdated to 16 billion years, with 10^{11} galaxies and 10^{11} stars per galaxy, our insignificant planet, a mere 4 billion years old, must have produced life early, as fossilized bacteria have been found at Warrawoona, Australia, in rocks believed to have been formed 3.5 billion years ago. Organic molecules such as nucleic acids, carbohydrates, and amino acids have been synthesized in conditions simulating those thought to have existed soon after the formation of the planet, but the path from such "spontaneous generation" to a living organism involved many giant steps. Strands of ribonucleic acid (RNA) can be replicated by autocatalysis, but linkage to polypeptide synthesis required an input of energy. Direct capture of sunlight could drive uphill reactions but limited activity to days. The use of radiant energy to split water, however, produced oxygen, which could then be used to metabolize stored molecules at night. The critical step for the existence of an "organism" must have been the production of hydrophilic-lipophilic molecules to form a bubble, a membrane to enclose a particular group of self-replicators, creating a fundamental barrier between Self and not-Self. From then onwards competitive evolution could take off, and from that primeval bubble life has proceeded in an unbroken sequence.

Living protozoa were first detected with a hand lens in the early seventeenth century, but the development of the compound microscope (1690–1710) encouraged the examination of thin sections of dead tissue. The term *cell* was coined in 1665 by Hooke, who likened the appearance of sections of cork to rows of monks' cells in a monastery. Histology thus became the study of inert fixed tissues, with limited resolution, so that organelles appeared to be suspended in a homogeneous cytoplasm. Electrophysiological experiments with microelectrodes sustained the view that the whole cytoplasm was a uniform conducting medium, with no potential gradients other than that across the membrane separating the cell from the external medium. Electron microscopy and other procedures, however, have revealed a great complexity of intracellular architecture, with individual organelles separated from the cytoplasm by membranes, opening the possibility that there is a voltage gradient between, say, the interior of the sarcoplasmic reticulum and its surroundings, though techniques for measuring such a gradient are still to be developed.

Histological sections gave a false impression of the stable and static nature of bodily tissues. Preparations of cardiac muscle looked much the same, whether taken from a man or a mouse. From a photograph of an airplane in the sky one could not deduce that it stayed aloft only because it was travelling at great speed. Research in biochemistry and molecular biology has shown that cellular components are being continuously synthesized and recycled in accordance with a multiplicity of instructions in response to ever-changing needs.

SIGNALLING BETWEEN CELLS

Unicellular organisms can be extremely complex, subdivided into compartments performing specialized functions, but the integrity of the frontier has always to be maintained. Ions and small molecules can be admitted or extruded only through carefully gated pores, and larger materials pass in (pinocytosis) or out (exocytosis) encased in envelopes that merge with the surface membrane (omegas). The functions of the different parts of the cell must be coordinated by internal messengers, but how these are translocated to their appropriate destinations is unresolved. Contraction, secretion, and other activities are associated with increases of cytoplasmic calcium concentration $[Ca]_i$ above its "resting" low level of 10^{-7} M, detected by light-emitting probes. Although such probes can record the timing of $[Ca]_i$ changes in relation to other electrical or mechanical events, the exact location of release and uptake of the calcium within the cell is beyond resolution. Similarly, cyclic adenosine monophosphate (cAMP) acts as an intracellular messenger for more than one function, and it is probable that specific structured pathways exist for transmitting messages from a source in one position in a cell to a destination in another, so that the same messenger can be employed for the coordination of different functions without confusion.

Living organisms in their external encounters are confronted with three fundamental questions: Can I eat it? Can it eat me? Can I mate with it? Recognition of prey, predator, or mate is the basis for survival. A single cell, however complex, cannot increase its volume-to-surface ratio beyond the point at which the rate of diffusion of gases and nutrients to its interior exceeds its metabolic requirement. The whole organism has to divide into two smaller independent cells. Another giant step in evolution was taken when two cells, instead of separating, remained in communication together as a single organism. Evolution is recapitulated in embryological development, and soon after fertilization dividing cells are still faced with the decision whether to continue together or separate into identical twins.

Multicellular plants appeared about 1200 million years ago, metazoa about 800 to 650 million years ago. A metazoon can be regarded as no different in principle from a protozoon. In the latter, particular functions are coordinated between separate parts of the cell; in the former, between separate, but coherent, cells. As the number of cells increased, so did the complexity of transmitting the signals needed to coordinate interrelated functions. Simple diffusion of molecules being too slow, an internal transport system was required. Even this became inadequate for the control of locomotion, and another giant step was the development of signals along the *surface* of specialized cells by the creation of electromagnetic fields. Concentration differences of ions between the inside and outside provided the field, but signals were still dependent upon temporary field-cancellation by diffusion of ions through selective channels, and the maximum velocity of conduction in even the largest myelinated nerves does not exceed 120 m.s^{-1}. Communication via optical fibers is so efficient and evolution has so often anticipated human invention, that one may wonder why some biological equivalent of the optical fiber never emerged. The compound eye of the housefly may be regarded a "near miss," providing a response usually fast enough to defeat a swatting hand.

Although each cell in the body performs a special function, it still contains all the information to create and control the entire organism. Differentiation is selection. To produce the total number of cells in a human would require the original zygote to divide, and its products to divide, through about forty-five generations, but many tissues, including the myocardium, lose the capacity for mitosis soon after their final differentiation. From birth the number of myocardial cells is fixed, and they can only increase in size. Hypertrophy can occur, but the width of cells is limited, not increasing beyond a diameter of 20 microns, because otherwise, adequate oxygen would not reach the cell interior. Myocardial cells increase in length by adding sarcomeres. There is no loss of capacity to divide by mitochondria, however, which increase in number, so that the ratio of mitochondrial volume to whole-cell volume is conserved. Reproduction of mitochondria is independent of cell reproduction, since they are derived directly from the cytoplasm of the maternal ovum.

The external signals which switch on myocardial hypertrophy, and mitochondrial hyperplasia, have been the topic of much experiment and discussion. Although adrenergic stimuli can induce hypertrophy (as can thyroid hormone) in association with raised intracellular cAMP, hypertrophy is still induced by increased intracardiac pressure after blockade of adrenoceptors or elimination of sympathetic drive by chemical sympathectomy (24,25). Angiotensin receptors may also play a role, since angiotensin enzyme (ACE) inhibitors restrict remodeling of the myocardium after infarction.

New signalling systems co-ordinating the functions of the body as a whole continue to be revealed at frequent intervals, and highlight ignorance of overall control. All cells have labels which enable the immune system to distinguish those which belong from invaders which do not, and mis-labelling can be induced with destructive consequences. Individual tissues have their own labels (receptors) which permit them to be targeted by messengers from elsewhere. Many cells may be uniquely labelled and recognizable, for example, by nerves growing towards their appropriate destinations. Conversely, unless nerve endings *receive* their appropriate signals they self-destruct.

Recognition of the remarkable complexity of signalling between cells encourages a fresh approach to pharmacology. Drugs imitate or block natural messages. Since, however, the number and function of messengers is certainly much greater than is known at present, caution is required both in interpreting drug action and anticipating side effects. Such ignorance has led to some surprises, in that an unexpected drug effect has led to the discovery of a natural control mechanism. Drugs introduced for one purpose have often proved useful for another. Amiodarone, used initially to control angina by antisympathetic action, became the most effective antiarrhythmic

agent; mexilitine was originally synthesized as a possible antiepileptic; ACE inhibitors not only reduce angiotensin II but increase bradykinin; thalidomide, a "safe" hypnotic, had disastrous effects on embryos.

INTRACARDIAC NERVOUS SYSTEM

The main control system that takes important decisions is, of course, the brain, and within the central nervous system (CNS) consciousness may be regarded as the part that excludes all information except that requiring an immediate response. The commander-in-chief must be protected from less urgent decisions by his subordinates, who carry out prearranged programs. Most of the information pouring into the brain from all over the body never reaches consciousness; indeed, the function of vast numbers of small afferent nerves from skeletal muscles is still unknown. Whole organs, notably the gastrointestinal tract, have their own nervous system with reflex arcs controlling mechanical and secretory functions locally, while the sympathetic and parasympathetic outflows exert a modulating influence only. In preparations of Thiry-Vella loops in conscious dogs, in which fluid propulsion was measured quantitatively while input and output pressures were varied independently, it was found that fluid "output" was proportional to input pressure (as in Starling's law of the heart) and that it was well maintained as outflow pressure was raised. Propulsive contractions occurred in bursts, and the frequency of such bursts also increased as input pressure was raised (26). The heart may likewise contain an internal nervous network, operating autonomously to coordinate appropriate timing of contractions in atria and ventricles locally, while the baroreceptor arc controls frequency overall.

Experimental support for this view was provided recently by Edwards et al. (27), who described three types of ganglion cell in the cardiac plexus of guinea-pig hearts. 1) S-cells, with a somatic action potential followed by a brief after-hyperpolarization. Few of these received synaptic input from the vagus, but all received a local synaptic input. 2) P-cells, with a somatic action potential followed by a prolonged after-hyperpolarization. None received a vagal input, nor did they respond to local input. 3) SAH cells receiving synaptic input from both vagus and local sources.

"S-cells had smaller somata than either of the other cell types and they were invariably monopolar. Since these cells appeared not to receive a synaptic input from the vago-sympathetic trunk, it would seem possible that these cells are mainly influenced *by synaptic projections that arise within the heart itself.* SAH cells serve to relay information from the CNS to the heart. They also appeared to receive projections that originated from within the heart. Some form of local handling of neuronal information might occur. We suggest that this may also occur in the cardiac plexus. A number of other observations suggest that *peripheral intrinsic circuits exist in the heart,* e.g. persistence of synaptic boutons after degeneration of sectioned vagal and sympathetic preganglionic projections" (27).

As in skeletal muscle the function of many afferent nerves reaching the CNS from the heart is still unknown. What is the physiological stimulus to nerves in the left ventricle excited by veratridine (von Bezold), which instigates reflex vagal activity? Are they responsible for the bradycardia often observed after posterior infarction? How and why does ischemia excite pain fibers? Ischemia may be detected electrocardiographically in the absence of pain, and pain can occur without objective evidence of ischemic damage. How does coronary flow keep pace with oxygen demand? Adenosine may be involved, but are there other controls? Atrial natriuretic peptide (ANP) is released in response to atrial distension, but is its function solely to control sodium reabsorption in the kidney, or does it influence cardiac activity also?

SUMMARY

Some gaps in current medical expertise have to be accepted openly and with humility.

Cot deaths (also termed *crib deaths* or *sudden infant death syndrome*) and sudden cardiac deaths may involve the failure of control systems not as yet understood. Uncertainty about much of the external control of coordinated contraction in the heart and about the cellular mechanisms by which messages are translated reinforces the importance of exploiting the knowledge that does exist. There are only four possible sources of net charge transfer, based on the concentration gradients of Na, Ca, K, and Cl ions, by passive diffusion through selective channels, or energy-coupled transport. Do all the numerous currents detected by experimental exercises on isolated single cardiac cells or patches of membrane have a real physiological function in whole human beating hearts?

A systematic study of the electrophysiological and other effects, on the SAN, atrium, AVN, His-Purkinje system, and ventricle of normal isolated functioning mammalian cardiac preparations, of drugs known to have antiarrhythmic properties in humans revealed that they all possessed one or more of five actions that could explain their antiarrhythmic efficacy. Furthermore, new antiarrhythmic agents subsequently introduced also possessed one or more of the same fundamental actions. From this factual experimental evidence, widely repeated and confirmed elsewhere, it was concluded that there is a limited number of interventions capable of preventing or correcting abnormal cardiac rhythm. As knowledge of cardiac control mechanisms expands additional antiarrhythmic actions may emerge, but at present this possibility remains in the realm of speculation.

REFERENCES

1. Vaughan Williams EM. The action of quinidine, acetylcholine and anaphylaxis interpreted from simultaneous records of contractions and intracellular potentials in the heart. *Scientific Basis of Medicine Annual Reviews.* London: Athlone Press, 1961; 302–323.
2. Ling G, Gerrard RW. The normal resting potential of frog sartorius fibres. *J Cell Comp Physiol* 1949; 34:383–396.
3. Vaughan Williams EM. The mode of action of quinidine in isolated rabbit atria interpreted from intracellular records. *Br J Pharmacol* 1958; 13:276–287.

4. West TC, Amory DW. Single fiber recording of the effects of quinidine at atrial and pacemaker sites in the isolated right atrium of the rabbit. *J Pharm Exp Ther* 1960; 130:183–193.

5. Vaughan Williams EM. The mode of action of drugs on intestinal motility. *Pharmacol Rev* 1954; 6:159–190.

6. Fatt P, Katz B. The electrical properties of crustacean muscle fibres. *J Physiol* 1953; 120:171–204.

7. Vaughan Williams EM. The effect of changes in extracellular potassium concentration on the intracellular potentials of isolated rabbit atria. *J Physiol* 1959; 146:411–427.

8. Hodgkin AL, Horowitz P. The influence of potassium and chloride ions on the membrane potential of single muscle fibres. *J Physiol* 1959; 148:127–160

9. Noble D. A modification of the Hodgkin-Huxley equations applicable to Purkinje fibre action and pacemaker potentials. *J Physiol* 1962; 160:317–352.

10. Vaughan Williams EM, Szekeres L. A comparison of tests for antifibrillatory action. *Br J Pharmacol* 1961; 17:424–432.

11. Black JW, Stephenson JS. Pharmacology of a new adrenergic beta-receptor-blocking compound, NETHALIDE. *Lancet* 1962 (2); 311–314.

12. Vaughan Williams EM, Sekiya A. Prevention of arrhythmias due to cardiac glycosides by block of beta sympathetic receptors. *Lancet* 1963 (1); 420–421.

13. Gill EW, Vaughan Williams EM. Local anaesthetic activity of the beta-adrenoceptor antagonist, pronethalol. *Nature* 1964; 201:199.

14. Dohadwalla AN, Freedberg AS, Vaughan Williams EM. The relevance of beta-receptor blockade to ouabain-induced cardiac arrhythmias. *Br J Pharmacol* 1969; 36:257–267.

15. Papp JGy, Vaughan Williams EM. Effect of bretylium on intracellular cardiac action potentials in relation to its antiarrhythmic and local anaesthetic activity. *Br J Pharmacol* 1969; 37:380–390.

16. Freedberg AS, Papp JGy, Vaughan Williams EM. The effect of altered thyroid state on atrial intracellular potentials. *J Physiol* 1970; 207:357–370.

17. Singh BN, Vaughan Williams EM. The effect of amiodarone, a new antianginal drug, on cardiac muscle. *Br J Pharmacol* 1970; 39:657–668

18. Vaughan Williams EM. Classification of antiarrhythmic drugs. In: *Symposium on Cardiac Arrhythmias*, Ed. E Sandœ, E Flensted-Jensen, KH Olesen AB Astra. Sweden:Sœdertälje, 1970; 449–472.

19. Singh BN, Vaughan Williams EM. A fourth class of antiarrhythmic action? Effect of verapamil on ouabain toxicity, on atrial and ventricular intracellular potentials, and on other features of cardiac function. *Cardiovasc Res* 1972; 6:109–119.

20. Kobinger W, Lillie C, Pichler L. N-allyl derivative of clonidine, a substance with specific bradycardic action at a cardiac site. *Naunyn-Schmiedebergs Arch Pharmacol* 1979; 306:255–262.

21. Millar JS, Vaughan Williams EM. Anion antagonism—a fifth class of antiarrhythmic action? *Lancet* 1981 (1); 1291–1293.

22. Millar JS, Vaughan Williams EM. Pacemaker selectivity. Effect on rabbit atria of ionic environment and of alinidine, a possible anion antagonist. *Cardiovasc Res* 1981; 15:335–350.

23. Vaughan Williams EM. Antiarrhythmic action of specific bradycardic agents. *Eur Heart J* 1987; 8 (Suppl L):17–18.

24. Dennis P, Vaughan Williams EM. Hypoxic cardiac hypertrophy is not inhibited by cardioselective or nonselective beta adrenoceptor antagonists. *J Physiol* 1982; 324:365–374.

25. Vaughan Williams EM, Dukes ID. The absence of effect of chemical sympathectomy on cardiac hypertrophy induced by hypoxia in young rabbits. *Cardiovasc Res* 1983; 17:379–389.

26. Streeten DHP, Vaughan Williams EM. The influence of intraluminal pressure upon the transport of fluid through cannulated Thiry-Vella loops in dogs. *J Physiol* 1951; 112:1–21.

27. Edwards FR, Hirst GDS, Klemm MF, Steele PA. Different types of ganglion cell in the cardiac plexus of guinea pigs. *J Physiol* 1995; 486:453–472.

2

Class 1 Antiarrhythmic Action

E. M. Vaughan Williams

CONDUCTION

In 1837 Charles Wheatstone patented a device for improvements in giving signals and sound alarms in distant places by means of electric currents transmitted through metallic circuits. In the same year Samuel Morse demonstrated to a skeptical US Congress a magnetic telegraph, but it was not until 1843 that Congress appropriated 30,000 dollars for an experimental telegraph line between Washington and Baltimore. Telegraph poles soon proliferated throughout the world's empires, and a poet could report upon a royal malady with the immortal lines

Across the wires the electric message came,
He is no better; he is much the same.

Submarine cables soon followed (first trans-Atlantic cable 1858), with four particular requirements: (a) a conductive central core, surrounded by insulation with (b) high resistance and (c) low capacitance, all enclosed within (d) a strong outer sheath for protection against shark and storm.

Unsurprisingly these problems had already been solved biologically. Single neurones from the spinal cord send messages down their delicate axons for up to a meter. Longitudinal resistance is reduced by increasing the diameter of fibers in which fast communication is essential (to and from muscles and some sensory organs). Schwann cells wrap themselves round the axon several times, increasing radial resistance, but reducing capacitance, at each revolution (Fig. 1) The fourth requirement is satisfied by the bundling of hundreds of fibers, large and small, inside a fibrous sheath, travelling parallel with arteries and veins. Even the signalling system anticipated the Morse code as a series of short on/off impulses.

The inward sodium current, which depolarizes the axon, is the integral of that passing through thousands of individual channels traversing the membrane. In myelinated nerve these are clustered at the nodes at a density of about 2000 per μm^2, with hardly any along the internodal regions (1). The high resistance (R_m) across the Schwann cell ensures that passive spread of capacitative current from inactive nodes towards an active node depolarizes not only the next node to threshold, but two or more beyond that, providing a high safety factor for "saltatory conduction."

At the negative resting potential the sodium channel is nonconducting (resting state, R). When the membrane is depolarized by about 15 mV by flow of capacitative current into the approaching active region, the channel may change into conducting mode (active state, A) and sodium ions flow in to depolarize the membrane further and increase the probability of neighboring channels opening. There is positive feedback, and sodium current soars to a maximum with explosive effect. The channels then flip into a depolarized but nonconducting mode (inactive state, I). Potassium channels open, meanwhile, in response to the depolarization (delayed rectification) and the axon rapidly repolarizes.

"Sodium channels are remarkably efficient in ion conductance. Voltage-clamp currents in squid giant axons or frog muscle fibers gave a unit conductance of 2.5 to 8.6 pS. Fourier transform methods yielded estimates of 4.1 to

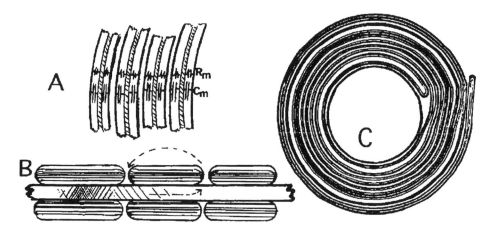

FIG 1. The Schwann cell encircles the axon many times **C**, ensuring that at each revolution **A** the radial resistance increases ($R_m = r_1 + r_2 + \dots r_n$) and capacitance decreases ($1/C_m = 1/c_1 + 1/c_2 + \dots 1/c_n$). **B** The high resistance ensures that the space constant (λ = distance along the axon, from the point of entry of direct current, at which V falls to V/e) is large enough for depolarization to reach threshold at the next inactive node.

8.8 pS in squid giant axon and frog node of Ranvier. Direct measurement by single-channel recording gave an estimate of 12 to 18 pS. All these estimates imply physiological ion transfer rates of $>10^7$ ions/s, consistent with movement through a fixed pore or channel. The functional properties of excitable cells are determined to a substantial extent by the density and location of sodium channels in their cell surface membranes. Differences in sodium channel density ... show that sodium channels are strikingly concentrated in some regions.... Sodium channels in regions of high cell-surface density are partially or completely immobilized. This mechanism likely serves to maintain differential distribution of sodium channels once they have been established" (1).

The fixation of clusters of sodium channels in specific locations suggests that other relevant structures, sodium pumps, for example, are also located close to them. Clustering of the channels facilitates positive feedback.

GAP JUNCTIONS

Skeletal muscles do not contract unless instructed by the CNS (central nervous system), and apparent synchronization in a sustained contraction is imposed by nerves, there being no electrical contact between neighbors. In contrast, the heart, though its performance is modulated by autonomic nerves and circulating hormones, is basically autonomous. The SAN (sinoatrial node) initiates depolarization, which then travels from cell to cell through electrically conducting gap junctions. In cultured myocytes small hexagonal structures appear on the surface, multiply, and assemble in patterns that recognize their counterparts on neighboring cells and unite to form multiple connecting channels—the gap junctions—wide enough to pass ions and small molecules (Fig. 2). Thereafter the contiguous cells beat in synchrony. In a growing embryo the cells are presumably always in contact, and the junctions develop quickly, as the primitive heart-tube pumps as an organ from a very early stage.

The surface membrane has a very low resistance compared with that of myelin, and so the space constant is very short. A line of contiguous cells conducts, as in an axon, by activation of sodium channels by discharge of membrane capacity into the approaching active region, but the saltation is only from

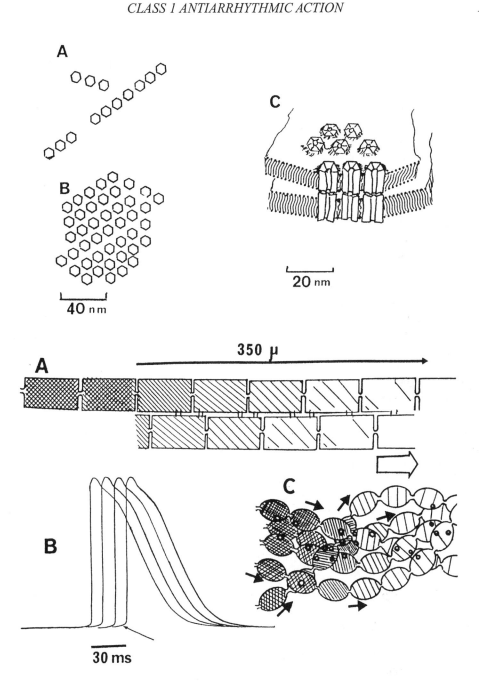

FIG 2. *Upper section.* **A, B.** Formations on cell surface. **C.** Union of surface structures to form gap junctions. *Lower section.* **A** Cell to cell conduction through gap junctions causes **B** successive action potentials to arise (for clarity every tenth potential only is displayed). **A and C** . Side-to-side connections through gap junctions between parallel neighbors synchronize depolarizations and unify the wave-front.

cluster to cluster of channels along the cell surface, instead of from node to node. Gap junctions occur between cells not only end to end, but also from side to side, thus synchronizing depolarization as a broad and sharply defined wave-front, without salients or retreats, reducing the probability of reentry. The very short space constant ensures a sharp demarcation between active and inactive areas, so that a large synchronous current flows across the border, sufficient to create potential differences at the body surface, recordable as the ECG. Every organ could, therefore, if it had the means to recognize it, receive a synchronization signal from the heart.

The supposition that the density and distribution of gap junctions determine the pathway of myocardial conduction has received direct experimental support. Connexin-43 has been identified as the principal structural component, and fluorescent antibodies raised against one of its intracellular loops have been used to label the junctions *in situ*. "Gap junctions within the intercalated disk in the human heart are typically organized as a peripheral ring of larger junctions, with smaller junctions centrally. They are in low quantity in the specialized myocytes of the SA (sinoatrial) and AV (atrioventricular) nodes" (2). Immunofluorescent labelling of connexin-43 confirms that they are numerous at the periphery of the guinea pig SAN and atrial fibers, but scarce as they penetrate the center of the node. Outward conduction is easy, but inward conduction is slow, with the result that an inward travelling action potential may not arrest the pacemaking of the innermost cells (3).

"The pattern of gap junction distribution in normal myocardium is believed to be one major determinant of the anisotropic pattern of electrical propagation, by which conduction parallel to the fiber orientation is up to four times more rapid than transverse to it. Measurements of conduction using voltage-sensitive dyes have demonstrated the importance of lateral connections for homogeneous propagation of the wave-front" (2). Gap junctions are *absent* from an infarct scar, which

cannot, therefore, contribute to conduction through it. "The mechanism of the most frequently encountered clinical ventricular tachycardia involves reentry localized to the *surviving* border of healed myocardial infarcts in man" (2). In nonischemic hypertrophied myocardium there is a 40% reduction of connexin-43 per unit volume of muscle. If the stimulus to hypertrophy does not switch on gap-junction production, are there other controls? In rats angiotensin II shortens APD (action potential duration) and decreases gap-junction conductance, and ACE inhibitors increase it (4). The latter effect could be antiarrhythmic by increasing conduction velocity of an action potential, which would now arrive at a still refractory cell, which had previously had time to recover excitability.

Objective evidence of gap-junction abnormality as an arrhythmogenic risk factor in heart disease was obtained by Sepp et al. In hearts from patients dying with hypertrophic cardiomyopathy gap junctions were labelled by immunohistochemical methods for presence of connexin-43. "Gap junction abnormalities included: 1) random distribution of gap junctions over the surface of myocytes, rather than localization to intercalated discs; 2) abundant side-to-side gap junction connections between adjacent myocytes; 3) formation of abnormally shaped gap junctions. *Circles of myocytes continuously interconnected* by gap junctions were also observed. The remodelling of gap junction distribution may underlie the formation of an arrhythmogenic substrate (5).

In experimental studies of antiarrhythmic actions two main strategies have been adopted. Abnormalities of cardiac rhythm can be induced in animals by a variety of procedures (use of aconitine, digitalis, or barium; programmed stimulation; coronary ligation, etc.), and the effects of drugs investigated empirically. Unfortunately such procedures do not produce consistent results, and even if they did, doubt would still remain concerning the relevance of such artifacts to human cardiac arrhythmias. In this field the only important target is man. An alternative approach is to study the detailed electrophysiological and pharmacological actions of known antiarrhythmic agents in the hope of finding some common properties to which their efficacy could be attributed and that might throw some light on the probable origin of the arrhythmias themselves.

ALTERNATIVE EXPLANATION FOR ANTIARRHYTHMIC ACTION

The conclusion from such studies, that quinidine and several other compounds including lidocaine and procainamide exerted their antiarrhythmic effect by restricting inward sodium current, was not universally accepted, and Bigger and Mandel (6) concluded from their own work "the results caused us to reject the hypothesis that lidocaine exerts electrophysiological effects essentially like those of procainamide or quinidine." Davis and Temte (7), discussing the effects of lidocaine and diphenylhydantoin (DPH), concluded that "a reduction in rising velocity is not a necessary feature for antiarrhythmic activity." On the contrary, they concluded that "the most significant effect of lidocaine with regard to its antiarrhythmic action is the prevention of decremental conduction in Purkinje fibers." They considered that DPH and propranolol acted in the same way as lidocaine. Basset and Hoffman (8) also took the view that "DPH and lidocaine may abolish a reentrant rhythm by *improving* conduction. DPH either *increases* or does not substantially alter membrane responsiveness and conduction velocity." There are two reasons why it might be concluded that therapeutic concentrations of lidocaine would be insufficient to reduce maximum velocity of depolarization V_{max}. The first is that if the extracellular potassium concentration is low (and the above authors employed a $[K^+]_o$ of 2.7 mmol/liter, little more than half the normal plasma level), the resting potential is hyperpolarized and the depressant effect on V_{max} counteracted (9). The second is that lidocaine delays recovery from inactivation, but for a brief period only, so that if the interval between action potentials is long, no effect on V_{max} will be seen.

Both these points are of clinical significance, because lidocaine may be ineffective in patients in whom, perhaps as a consequence of diuretic therapy, serum potassium has fallen. Secondly, even when no effect on His-Ventricle or HV conduction time or QRS width is apparent in sinus rhythm, the Class 1 effect of lidocaine could nevertheless be responsible for depressing or eliminating conduction of *premature* ventricular beats or for slowing a ventricular tachycardia. In a study of experimentally induced ventricular arrhythmias in the late myocardial infarction period in dogs, it was found that at the time when lidocaine and DPH exerted an antiarrhythmic action, the effect of both drugs was to slow conduction. It was concluded that "there is currently no basis to substantiate the concept that both lidocaine and DPH can abolish reentrant rhythms by improving conduction in the reentrant pathway" (10).

Another complicating factor is that some drugs, notably lidocaine and mexiletine, *shorten* Action Potential Duration or APD in the ventricular conduction pathway, especially in the preterminal Purkinje fibers in which APD is normally much longer than in the His bundle or ventricular myocardium.

SUBDIVISION OF CLASS 1 ANTIARRHYTHMIC AGENTS

By definition Class 1 drugs share the property of restricting fast inward sodium current but may differ in other respects.

1. Some compounds, notably quinidine and disopyramide, are anticholinergic. Apart from noncardiac complications of antimuscarinic actions, there are several cardiac consequences. In a patient with substantial vagal activity, sinus tachycardia may follow administration of one of these compounds; or, more seriously, an atrial fibrillation or supraventricular tachycardia that was previously innocuous because impulses were blocked at the AV node may

develop into ventricular tachycardia when AV conduction is improved by blockade of vagal tone. Second, the fact that the drugs depress conduction through their Class 1 action may be masked by the simultaneous improvement of conduction by the anticholinergic effect, with no net change in AV conduction time. In a patient with little vagal background tone, however, and with depressed AV conduction, the full Class 1 depression by quinidine or disopyramide will be revealed, and serious AV block may occur (11).

2. Distribution into the CNS varies greatly among Class 1 drugs, and more information is required concerning the extent to which this is influenced by fat solubility, negative log of acid dissociation constant pKa, and molecular size. Certainly dizziness and other CNS effects, even convulsions, limit the use of several Class 1 agents, including lidocaine, mexiletine, and lorcainide.

3. Metabolism and rates of excretion vary enormously, both factors influencing elimination. Rapid metabolism by the liver makes oral administration of lidocaine impractical, since its half-life may be only 30 minutes in some subjects. At the other extreme aprindine has an elimination half-life of two to five days. The half-lives and volumes of distribution of nine Class 1 drugs are shown in Table 1 (12).

4. Some compounds, especially lidocaine, mexiletine, and tocainide, shorten APD, as already discussed.

5. There are wide variations in the rapidity with which the drugs become attached to and released from the sodium channels.

The *clinical* differences between Class 1 drugs necessitated their subdivision. The older compounds, quinidine, procainamide, and disopyramide, were classified as group 1a. Lidocaine, mexiletine, and tocainide (group 1b) shorten APD *in vitro* and monophasic action potentials (MAPs) and J-T intervals *in vivo*. They do not lengthen H-V conduction time or widen QRS in sinus rhythm, but prolong effective refractory period (ERP) as measured by programmed stimulation (a premature stimulus, S2, interpolated after every nth pacing stimulus, S1). The most recently introduced Class 1 drugs, including flecainide, encainide, and lorcainide (group 1c), have little effect on APD or J-T interval and do not prolong ERP on programmed stimulation, but increase H-V conduction time and widen QRS even in sinus rhythm. Quinidine, procainamide, and disopyramide are intermediate, moderately prolonging ERP and J-T interval and widening QRS at high concentration (indicating possible overdosage). The anticholinergic and *APD-lengthening* effects of quinidine are additional properties unrelated to its Class 1 action. Quinidine delays repolarization by restriction of outward potassium current (13). The differences between the Class 1a, 1b, and 1c groups are shown in Table 2.

KINETICS OF ATTACHMENT AND DETACHMENT OF CLASS 1 DRUGS

In view of these clinical electrophysiological differences some physicians were reluctant to accept that the antiarrhythmic effects could be attributed to a common action in blocking sodium channels. For his doctoral thesis TJ Campbell developed an elegant method of printing out columns, the height of which was proportional to V_{max} recorded with microelectrodes from guinea pig ventricle, paced at various frequencies. In control solution V_{max} was stable during a train of action potentials at 3 Hz. Stimulation was arrested

TABLE 1. *Elimination half-lives and volumes of distribution*

	Elimination of half-life (h)	Volume of distribution (liters/kg)
Lidocaine	1.8	1.6
Mexiletine	13.0	6.6
Tocainide	13.0	2.8
Disopyramide	7.0	0.5
Quinidine	6.3	2.5
Procainamide	3.0	2.9
Flecainide	14.0	
Lorcainide	7.7	7.9
Aprindine	50.0	4.0

TABLE 2. *Clinical subdivision of Class 1 antiarrhythmic drugs*

Effect on	Group B	A	C
	Lidocaine	Quinidine	Propafenone
	Mexiletine	Procainamide	Moricizine
	Tocainide	Disopyramide	Flecainide
1. QRS	None in sinus rhythm	Widen at high concentration	Widen at low concentration
2. Conduction	None in sinus rhythm	Slowed at high concentration	Slowed at low concentration
3. ERP	Lengthened relative to APD	Lengthened absolutely and relative to APD	Very little change
4. J-T Interval	Shortened	Lengthened at high concentration	No change

and the preparation was exposed to flecainide 5 μM for several minutes. When stimulation recommenced, V_{max} in response to the first stimulus was unchanged, but declined progressively to a steady state, 40% slower after 32 stimuli, a rate of onset to steady state of 0.03 per beat. Stimulation was arrested, and the concentration of flecainide was doubled. When stimulation recommenced after ten minutes, V_{max} in response to the first stimulus was again almost normal, but declined to steady state after only twenty-three stimuli, and V_{max} was 60% slower. Thus both steady-state block and the rate of onset of block were concentration dependent (Fig. 3). From similar experiments in which steady state block was obtained in repeated trains of stimuli, stimulation being restarted (in continued presence of the drug) after increasing intervals, the rate of recovery from block (offset time constant, τ-rec) could also be determined

and was found to be independent of concentration (14).

The kinetics of many drugs with Class 1 action were studied and results for nine of them have been summarized in Table 3. An attempt was made in each case to choose a concentration that would produce approximately 50% steady-state block, and the actual degree of block observed is shown in the first column and the concentration required to produce it in the second. The fraction of steady block achieved per beat is in the third column, and the offset time constant in the fourth The remaining columns show physical data: pKa, fat solubility (log of the octanol/water partition coefficient), and the molecular weight.

There was a hundredfold range of potency, and rapidity of onset varied from less than one beat to fifty. There was no correlation between activity and pKa or fat solubility, but a good fit between offset time constant and

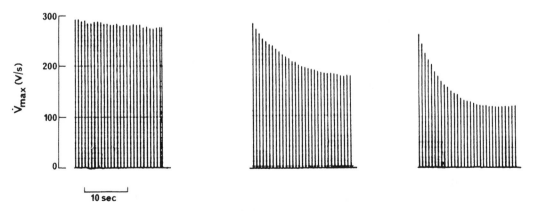

FIG. 3. *Left:* V_{max} is well maintained during a train of stimuli at 3 Hz. *Center:* in presence of 5 μM and, *Right,* 10 μM flecainide, steady block and rate of onset are concentration dependent.

TABLE 3. *Onset/offset kinetics of frequency-dependent block by Class 1 antiarrhythmic drugs. Physical characteristics*

Drug	% Block	Concentration microM	Block per beat	Tau-rec sec.	pKa	logp	M.W.
Lidocaine	40.4	200	>0.6	0.3	7.85	2.76	234
Mexiletine	49.2	20	>0.6	0.471	9.3	1.3	179
Tocainide	45.2	300	0.277	1.1	7.8	0.8	190
Disopyramide	50.9	100	0.133	2.2	8.4	1.8	339
Quinidine	52.5	30	0.068	4.7	8.8	3.6	324
Procainamide	41.0	180	0.055	2.3	9.2	0.8	236
Flecainide	50.7	5	0.029	15.5	9.3	1.15	408
Encainide	50.0	2	0.025	20.3	10.2	0.91	358
Lorcainide	45.0	2	0.022	13.2	9.44	4.16	371

molecular weight ($r = 0.94$, $p > .001$), which suggests that once the drug is attached to the channel, it may be held by additional short-range van der Waals forces.

The table, based upon experimental studies *in vitro*, groups the drugs into the same three categories based on clinical evidence. Lidocaine, mexiletine, and tocainide—with the fastest onset and offset kinetics—represent Class 1b; and flecainide, encainide and lorcainide—with the slowest—represent 1c. Quinidine, procainamide, and disopyramide—with intermediate kinetics—represent 1a. A continuous spectrum of frequencies of visible light is split by the eye into red, green, and blue; by analogy there is an extended spectrum of onset/offset kinetics of drugs with Class 1 action, as shown in Fig. 4.

Lack of effect on quiescent tissue suggests that a drug does not combine with a sodium channel in its resting state and that a site for attachment is revealed only after depolarization (an old proposition elaborated more recently as the "guarded receptor" model). The 1c compounds dissociate so slowly that at steady state they eliminate a fraction of sodium channels permanently so long as the drug is present. Sodium current is reduced, conduction velocity is slow, and QRS, wide. The muscle depends for its continued viability on the fraction of channels *free of drug*, which therefore recover from inactivation at the normal time after repolarization. Thus the cell, if it responds at all, will *respond without delay*, and there will be *no change in*

refractory period. In contrast, the 1b compounds block the channels so rapidly that most are blocked by the end of the action potential, and *ERP is prolonged.* Soon after repolarization, dissociation proceeds so quickly that by the end of a *normal* diastole (sinus rhythm) nearly all channels are again drug free, sodium current is normal, and conduction velocity and QRS (determined by the upstroke of the action potential) are unchanged. If diastole is short, however, the channels will still have drug attached, and conduction of a premature beat will be delayed or blocked.

The onset-offset kinetics thus explain most of the differences between the clinical behavior of the Class 1a, 1b, and 1c drugs. Anticholinergic action and prolongation of APD are accounted for by the *additional* properties of some drugs, unrelated to action on sodium channels.

Kidwell et al. (15) employed lidocaine, procainamide and flecainide to treat patients with ventricular tachycardia, in an attempt to assess whether the different onset time constants reported *in vitro* were of relevance to the time taken to establish a change in cycle length of the Ventricular Tachycardia or VT. Use-dependent prolongation of VT cycle length during treatment with Type I antiarrhythmic drugs was observed in humans. The estimated time constants for the use-dependent prolongation of VT cycle length by the three test drugs are similar to their reported *in vitro* time constants for use-dependent sodium channel blockade.

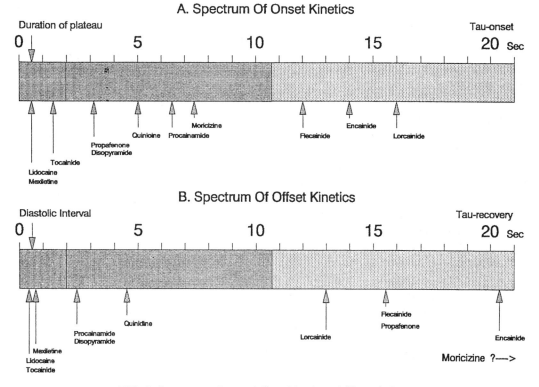

FIG. 4 Spectrum of onset/offset kinetics of Class 1 drugs.

THE CARDIAC ARRHYTHMIA SUPPRESSION TRIAL

The US National Heart, Lung and Blood Institute sponsored a pilot study in 1986 to determine the feasibility of conducting a full-scale trial to test the hypothesis that *reducing* ventricular arrhythmias after acute myocardial infarction (AMI) decreases mortality risk (16). The results of that study were published two years later (17). During a double-blind drug and dose selection phase, investigators were permitted to change drug or dosage to achieve >70% *suppression* of ventricular premature contraction (VPC) frequency and ≥90% *suppression* of runs of VPC. The trial itself was initiated as the Cardiac Arrhythmia *Suppression* Trial (CAST), a deviation from the *reduction* at first envisaged. The frequency of VPCs is so variable that a reduction of less than 70% might not be statistically significant.

A preliminary report published a year and a half later (18) concluded that "neither encainide nor flecainide should be used in the treatment of patients with asymptomatic or minimally symptomatic ventricular arrhythmia after myocardial infarction, even though these drugs may be effective in suppressing ventricular arrhythmia." The trial monitors had found that out of 734 patients taking these drugs, 33 had died, compared with 9 deaths out of 725 in patients on placebo. Encainide and flecainide are both in the Class 1c category. Moricizine, the third drug employed in the trial, came from the USSR and less was known about its pharmacology. The trial, with some modification, continued with moricizine, as CAST 2. Meanwhile, from a careful study of the data available, it was concluded that moricizine also should have been included in the Class 1c category (19), in spite of previous statements that it had Class 1a and 1b properties. Eventually CAST 2 was also discontinued because "1) patients treated with moricizine had an excess mortality rate during the 1st 2 weeks of exposure to the drug; and 2) there appeared to be little chance of showing a long-term survival benefit from treatment with moricizine" (20).

The impact of CAST was remarkable, provoking many an editorial, and a reluctance among some cardiologists to prescribe Class 1 drugs at all to post-AMI patients. It was assumed by some authors that most excess trial deaths were caused by arrhythmias. "Although its basis is not entirely clear, this unexpected result is best explained as the result of the induction of lethal arrhythmias (ie, a proarrhythmic effect)" (21). In contrast, Bigger (22) emphasized that the mechanism of the increased mortality in the patients treated with encainide or flecainide is not known, and gave several reasons for doubting that the effect was proarrhythmic. In some patients, at least, the drugs may have exacerbated heart failure, a view supported by Gottlieb (23). As already noted, Class 1c compounds remove a proportion of functional sodium channels. Long-term treatment with high doses implies that the heart is dependent on the remaining channels that are drug-free. If the resting potential falls, as it will in ischemic tissue, fewer channels recover from inactivation, and a vicious circle arises, until insufficient channels are available to sustain conduction, resulting in cardiac arrest.

Achievement of a >70% suppression of VPCs required a high dosage of 1c drugs in the trial, administered long term. If the original proposal of the pilot study (to test the hypothesis that reducing ventricular arrhythmias decreases mortality risk) had been adhered to, and if patients had been treated with lower doses of the drugs, *reducing* the frequency of VPCs to a minor, even statistically insignificant, extent, the result of the trial could have been different. Isolated VPCs, or short runs, are often symptomless and innocuous in themselves, but may be harbingers of more serious life-threatening ventricular tachycardia. Treatment with an antiarrhythmic drug, at a dosage sufficient to

reduce, but not suppress, VPCs could prevent progression to a lethal arrhythmia. This possibility was not excluded by CAST. Thus the important lesson to be learned from CAST was *not* that Class 1 drugs should *never* be administered to post-MI patients, but that the objective of therapy should not be the *suppression* of VPCs, but the administration of the lowest dose of an antiarrhythmic drug capable of controlling debilitating symptoms, with the possible bonus that such control might also avoid progression to a more serious and life-threatening arrhythmia.

EFFECTS OF INDIVIDUAL CLASS 1A AGENTS

Quinidine

During 80 years of use quinidine has not been subjected to the rigorous scrutiny of a prospective trial now compulsory for a new drug. A meta-analysis of six trials published between 1970 and 1984 involving 808 subjects revealed that the proportion of patients remaining in sinus rhythm 3, 6, and 12 months after electrical cardioversion was 69%, 58%, and 50%, respectively, in those receiving quinidine and 45%, 33%, and 25%, respectively, in the control group. The unadjusted total mortality, however, was 2.9% in the quinidine cohort, but only 0.8% in the controls. Quinidine treatment is more effective than no antiarrhythmic therapy in suppressing recurrence of atrial fibrillation, but appears to be associated with increased total mortality (24).

Atrial Arrhythmias

Quinidine is still widely prescribed for atrial arrhythmias, although other effective and better tolerated compounds are available. The additional action on outward K^+ current, prolonging APD, is advantageous in atrial muscle, with its short action potential, but the reverse frequency dependence of this effect makes it less useful at high rates. The anticholinergic effect, however, can be a disadvantage by relieving AV block and so increasing ventricular rate in atrial fibrillation. Physicians traditionally prescribing quinidine are likely to have their patients on digoxin already, so that the anticholinergic action on AV conduction would be nullified. Combination of quinidine and digoxin causes an increase in serum digoxin. Quinidine has several unpleasant noncardiac side effects, of which diarrhea is the most common. Being a stereoisomer of quinine, the drug may induce the well-known complication of antimalarial treatment, cinchonism, with CNS effects, headache, and tinnitus.

Some advantages of quinidine are good absorption on oral administration and compatibility with poor ventricular or renal function, because clinical concentrations have little negative inotropic action, and only 20% is excreted unchanged by the kidneys. The metabolites of quinidine, hydroxy-3S-quinidine, and of hydroquinidine, hydroxy-3S-hydroquinidine, have only half the activity of quinidine on V_{max}, but a greater, and frequency-independent, activity on APD (25).

Ventricular Arrhythmias

Quinidine depresses V_{max} and prolongs monophasic action potential duration (MAPD) in ventricular muscle also. In patients with overt right ventricular disease treated with quinidine or procainamide it was shown that the effect of both drugs on MAPD exhibited reverse frequency dependence (26). The positive frequency dependence of the effect on V_{max} and reverse frequency effect on APD tend to cancel each other, so that the net effect of quinidine on the ventricular effective refractory period (VERP) becomes frequency independent (27). These findings were confirmed by Nademanee et al.: "quinidine's effect on the action potential duration was attenuated at the pacing cycle length of 350 ms, and became negligible at 300 ms. Quinidine uniformly increased the ventricular refractory periods by about 9–11%, regardless

of the pacing cycle lengths and the action potential duration. In contrast to quinidine's effect on APD, the drug prolonged QRS duration in a use-dependent manner" (28).

Procainamide

Sodium channels in nerves differ from those in cardiac muscles, but in spite of a wide range of selectivity, the similarities are such that most local anesthetics have antiarrhythmic properties and vice versa. Procaine is split so rapidly by cholinesterase that it has a very short life in blood, but the derivative procainamide lasts a few hours and is sufficiently selective for cardiac sodium channels to be useful as an antiarrhythmic drug. Procainamide lacks anticholinergic effects, but like quinidine its metabolite loses activity on V_{max}, but has increased activity on APD. *N*-acetyl procainamide (NAPA) has itself been used as a Class 3 agent, and its presence may contribute to the antiarrhythmic action of procainamide. Long-term treatment is less safe than with quinidine, because of the induction of antinuclear antibodies and eventual development of the lupus syndrome. Nevertheless procainamide has been widely used for the treatment of atrial and ventricular arrhythmias. An unexplained observation with procainamide in patients with an accessory conduction pathway between atrium and ventricle is a preferential effect on anterograde conduction, greater than on retrograde conduction (29). Comparisons with other Class 1 drugs were not made.

As with quinidine, the onset/offset time constants for the effect of procainamide measured *in vivo* by recording MAPD were similar to those observed with microelectrodes *in vitro*. The rate constant for the onset of block in anesthetized dogs was frequency dependent; the offset time constant was 3.4 seconds. "These results support the relevance of basic concepts about antiarrhythmic drug actions on sodium channels for understanding drug effects *in vivo*" (30). The frequency dependence of the effect of procainamide on VERP has been confirmed in humans also (31). Although long-term treatment with procainamide is limited by risks of toxicity, it has proved useful for diagnostic purposes. "The conversion of inducible polymorphic ventricular tachycardia to uniform tachycardia after procainamide occurs almost exclusively in patients with coronary artery disease, previous MI, and abnormal ventricular function. This response may permit activation mapping of tachycardia, allowing the application of surgical or ablation techniques that otherwise would not be possible in such patients (32)." "The effect of procainamide on inducibility of ventricular tachycardia during programmed ventricular stimulation can be predicted *by the degree of drug-induced prolongation of the signal-averaged QRS complex*" (33).

The efficacy of Class 1 drugs in slowing depolarization may predict success as an antiarrhythmic, but can success as an antiarrhythmic predict longer survival? Goldstein et al. reexamined the CAST data, separating a group of patients whose arrhythmia was easy to suppress (>80% suppression of Ventricular Premature Depolarizations or VPDs on the first dose of the first drug) from a hard-to-suppress group (requiring increasing doses or change of drug). Their conclusion was that "ease of VPD suppression identifies a subgroup of post MI patients who have a low risk of arrhythmic death" (34).

Disopyramide

Disopyramide was selected from more than 500 compounds synthesized for a research program designed to develop a new antiarrhythmic drug as an alternative to quinidine or procainamide, the main agents available before 1962. In conventional screening tests for antiarrhythmic activity disopyramide suppressed aconitine-induced atrial arrhythmias in anesthetized dogs at a dose of 2.7 mg.kg^{-1} and electrically induced arrhythmias at a dose of 5.8 mg.kg^{-1}. The first electrophysiological study with microelectrodes revealed effects very similar to those of quinidine, a frequency-dependent depression of V_{max}, and a prolonga-

tion of APD (35). In anesthetized dogs effects of vagal stimulation were abolished by 5 mg/kg^{-1}, and cardiac output was reduced by 30%. In conscious dogs, however, cardiac output was well maintained and heart rate increased, indicating no interference with reflex responses to a negative inotropic action. A detailed review of the *in vitro* studies and clinical effects of disopyramide after twenty years of use was published in 1984 (36).

In humans disopyramide prolonged ERP of the atrium, but reduced that of the AV node, and it was suggested that the drug "might be indicated in the treatment of arrhythmias which are associated with some degree of AV block." Depression of AV conduction by disopyramide may be masked by its atropine-like action, and the net effect will depend on background vagal tone, which is likely to be low or absent in patients with poor ventricular function. If atropine is administered first, disopyramide always delays AV conduction (37). The anticholinergic action has sometimes caused urinary retention.

Disopyramide is well absorbed after oral administration and is removed both by renal excretion and metabolism, metabolites having little activity. It has been used extensively for the control of multiple ectopic beats, close-coupled extrasystoles, ventricular tachycardia, and arrhythmias involving an accessory pathway. The major limitation of disopyramide is its negative inotropic action. In patients with impaired ventricular function disopyramide increased pulmonary wedge pressure and left ventricular end diastolic pressure (LVEDP) and caused a depression of cardiac index by 28%. There was a 55% risk of inducing cardiac failure in subjects who had previously experienced an episode of such failure, and a 5% risk in those who had not (36).

EFFECTS OF
INDIVIDUAL CLASS 1B AGENTS

Lidocaine

Since its introduction in 1948 lidocaine has remained one of the most widely used local anesthetics. It was first employed as an antiarrhythmic during surgery, rapid elimination being an advantage. The very short duration of action made it unsuitable as an orally administered drug, but it is still often the antiarrhythmic of first choice in hospital. In sinus rhythm lidocaine has little effect on QRS or HV conduction time, and does not lengthen, or may shorten, the JT interval. In electrophysiological studies on cardiac tissues *in vitro* lidocaine prolonged ERP, but shortened APD, and it was claimed that it did not depress V_{max} except at higher than therapeutic concentrations. It was suggested that the antiarrhythmic action of lidocaine was fundamentally different from that of quinidine or procainamide, and was due to an increase in potassium conductance, which shortened APD, and *accelerated* conduction of premature action potentials by permitting them to "take off" from a fully repolarized membrane.

It was noted, however, that the experiments upon which this suggestion was based were undertaken at a very low potassium concentration (2.7 mM). When the bathing solution contained a potassium concentration similar to that of plasma, lidocaine, even at lower than therapeutic concentrations, depressed V_{max} in cardiac muscle as did other local anesthetics (9). It had long been realized that the toxicity of digitalis was greatly increased in hypokalemia, and when diuretics were introduced for the treatment of hypertension, cardiac arrhythmias associated with a fall in plasma potassium became a recognized risk. The inward rectifier current, I_{K1}, is activated on repolarization and stabilizes the resting potential during diastole. When the intracellular potential, E, becomes positive to E_K, the inward rectifier current falls, being negligible at the plateau potential. In hypokalemia, although the resting potential is more negative, it deviates further from E_K, so that fewer I_{K1} channels are open, and the resting potential is less stable. (See Chapter 1, Ref. 17) The hyperpolarized resting potential ensures that sodium channels are fully recovered from inactivation, and V_{max} is at its peak value. All Class 1 drugs are, therefore, less effective in low-potassium solutions. Conversely, in hyperkalemia, the resting potential is less negative, not all sodium channels may be available, and Class 1 drugs become more effective.

Lacking the APD-prolonging effects of Class 1a compounds, high concentrations of lidocaine are often required to produce a satisfactory antiarrhythmic effect, leading to CNS side effects, tremor, even convulsions, which set the chief limitation to the use of lidocaine, rather than a negative inotropic action. It was suggested that such high concentrations were responsible for APD-shortening by lidocaine, by blocking the persistent sodium current that helps to maintain the action potential plateau (38). "In Purkinje fibers sodium channels are not fully inactivated at the potential of the plateau." Coraboeuf and Deroubaix, in a study of canine Purkinje fibers, found that tetrodotoxin (TTX) shortened APD and stated "the effect is probably due to the fact that in Purkinje fibers the TTX-sensitive steady-state sodium current flows in a larger potential range." Whether this persistent current is through channels that have failed to inactivate, or through channels that open late, is still controversial. Experimental evidence that supports the view that lidocaine does block the persistent plateau current was published by Ju et al. (39), but since it was obtained in rat ventricular myocytes, its relevance is uncertain.

The most important and clinically relevant difference between lidocaine and the 1a drugs is the extreme rapidity of its onset/offset kinetics. In human atrial and ventricular myocytes isolated enzymatically from tissue removed during surgery, lidocaine blocked sodium channels in both their activated and inactivated state (40). There was no significant difference between effects on atrial and ventricular channels. Onset block had two time constants, 4.1 and 175 ms in atrium, 3.6 and 135 ms in ventricle. The recovery-time constant was 200 ms in the absence of drug, extended by 200 μM lidocaine to 700 ms, which explains why lidocaine prolongs ERP (response to a stimulus applied at the *begin-*

ning of diastole). Since, however, even in the presence of lidocaine recovery is fast enough to be almost complete by the end of a normal diastole in sinus rhythm, there is little effect on QRS or HV conduction time. In the atrium APD is shorter than in the ventricle, and diastole is correspondingly longer, allowing more time for the sodium channels to recover, which explains why lidocaine is less effective in atrial arrhythmias.

Shortening of APD is potentially proarrhythmic, so that the accelerated repolarization by lidocaine detracts from its antiarrhythmic efficacy. Hypoxia shortens APD, even in the presence of glucose and absence of any fall in ATP from its usual millimolar levels. In severe and prolonged ischemia, when ATP drops to micromolar levels, potassium channels open and an outward current (I_{K-ATP}) rapidly repolarizes the cell. In experiments on pigs after coronary ligation a concentration of lidocaine, which was insufficient to shorten APD in the nonischemic region, *increased* APD-shortening in the ischemic zone. Lidocaine also shortened the time to the appearance of arrhythmias, and *reduced* the threshold of stimulation required to induce fibrillation (41).

Mexiletine

The compound Kö 1173, later named mexiletine, had an anticonvulsant activity in animal experiments comparable to that of phenobarbitone and diphenylhydantoin (DPH). Since the latter had been used as an antiarrhythmic, it was unsurprising that mexiletine exhibited antiarrhythmic activity in animal models. It was reported to lack local anesthetic activity on the rabbit cornea. The first electrophysiological studies with microelectrodes on cardiac muscle revealed that it depressed V_{max} but did not prolong APD, and it shortened APD at higher concentrations (42). It was found to be twice as potent as procaine as a local anesthetic on desheathed frog nerve. It was concluded that its antiarrhythmic properties were due to a Class 1 action

similar to that of lidocaine. "This is not to imply that there are no differences between DPH, lidocaine and other drugs with Class 1 action, but if there are differences they may be due to extracardiac factors or to variations in relative sensitivity, rather than to any fundamental difference in their mode of action on the cardiac membrane" (42).

Since 1972 mexiletine became established as an antiarrhythmic drug with a profile similar to that of lidocaine. It was admitted that both lidocaine and mexiletine owed their antiarrhythmic properties to block of sodium channels, but since they had similar clinical effects (ERP prolonged, APD unchanged or shortened, QRS and HV conduction time unaltered in sinus rhythm), it was suggested that they should be listed separately as Class 1b compounds, while quinidine, procainamide and disopyramide were grouped as in Class 1a. Subsequently it was shown that mexiletine had fast onset/offset kinetics, confirming the relevance of these to the clinical characteristics. The main advantage of mexiletine over lidocaine was good oral availability and a longer elimination half-life, but CNS side effects (nausea, tremor) were similar.

The effect of 100 μM mexiletine in shortening APD in guinea pig papillary muscle was abolished by 50 μM glibenclamide, which blocks I_{K-ATP} channels (43,44). The possibility that mexiletine might be a potassium-channel opener in cardiac muscle was doubtful, however, because 100 μM is ten times the therapeutic concentration. In patch-clamp studies on guinea pig cardiac myocytes 40 μM mexiletine did not alter single-sodium-channel conductance but reduced the probability of channel opening, thus reducing peak average sodium current. "Mexiletine reduced the later occurrence of first openings. Additionally the late openings were reduced in a use-dependent way" (45). These findings support the view that the persistent sodium current is due to late activation of sodium channels, rather than to failure of open channels to inactivate, and that APD shortening by Class 1 drugs is due to abolition of such late openings.

Tocainide

Tocainide is structurally similar to mexiletine, having an -NH.CO- instead of -O.CH₃- between ring and side chain. In its electrophysiological and clinical actions and side effects, tocainide closely resembles mexiletine. The onset time constant of tocainide is the same as that of lidocaine, but the offset time constant, though short (1.5 s) is significantly longer.

EFFECTS OF
INDIVIDUAL CLASS 1C AGENTS

Flecainide

The first microelectrode electrophysiological studies of R818 (later named flecainide) were undertaken twenty years ago. They were reported to Riker Laboratories, but were not submitted for publication at the time because company takeover proceedings created doubt whether the drug would progress through clinical phases. The results, published later (46), indicated that flecainide was a Class 1 agent more potent than others currently available, a concentration of only 0.42 μM depressing V_{max} by 28% in rabbit atrial muscle. It had little negative inotropic action, and did not antagonize the positive inotropic effect of raised $[Ca]_o$. The spontaneous frequency of isolated atria was unaffected, but flecainide did prolong APD in atrial and ventricular myocardium. It was five times more potent than procaine as a local anesthetic on desheathed frog nerve, with a very prolonged action.

Flecainide widens QRS and prolongs QT only by the additional QRS width, so that JT interval is unchanged. The HV conduction time was increased. Flecainide thus exhibited clinical characteristics so different from those of the 1a and 1b drugs that, together with the now discontinued encainide, it had to be placed in a third category of sodium channel blockers, Class 1c.

In vitro flecainide exhibited little affinity for sodium channels in their resting state, and the onset/offset time constants of frequency-dependent block were very slow, 12 s and 15.5 s, respectively, confirming the relevance of kinetics to clinical effects.

The prolongation of atrial APD demonstrated *in vitro* was confirmed in humans with normal ventricles and in sinus rhythm, by a lengthening of ERP and of the atrial electrogram, which was maintained on pacing at higher frequencies (no reverse frequency dependence) (47). Excessive concentrations of flecainide (>1 mg.kg⁻¹) were proarrhythmic. The frequency-independent effect of flecainide on APD coupled with its Class 1 action presaged efficacy in atrial arrhythmias; this was borne out in practice. In 335 patients "with paroxysmal supraventricular tachyarrhythmias and without clinically significant heart disease, an intention-to-treat analysis showed that the probability of 12 months' safe and effective treatment of paroxysmal supraventricular tachycardia (SVT) was 93% with flecainide 100 mg/day, whereas in paroxysmal supraventricular fibrillation it was 77% with 200–300 mg/day" (48). Flecainide was also effective in patients with SVT associated with an accessory atrioventricular pathway, and a frequency-dependent *preferential* block of the accessory pathway was demonstrated, which was an "excellent marker of drug efficacy" (49). In a comparative study the frequency dependence of the effects of flecainide was greater in human isolated cardiac tissue (>dog>rabbit>guinea pig), and the positive frequency dependence of the prolongation of APD by flecainide and the reverse frequency dependence of the effect of quinidine were confirmed (50).

In the early nineties many papers on the effects of flecainide in different species were published; these confirmed the earlier findings, and attention was turned to attempts to identify the mechanism responsible for the prolongation of APD. In feline ventricular myocytes flecainide "10μM markedly reduced I_K elicited on depolarization steps to plateau voltages (+10 mV) and nearly completely blocked the 'tail currents' elicited on repolarization to –40 mV." Flecainide had no effect on the background potassium current $I_{K1}(51)$. In canine atrial tissue "flecainide 4.5 μM increased APD, with greater effect the faster the stimulation frequency. "The major time-dependent outward current in canine atrial cells, the transient outward current, I_{TO}, was reduced in a *rate-independent* fashion. Flecainide did not alter inward calcium current or the calcium-sensitive component of I_{TO}." It was concluded "the rate dependence of flecainide's action on APD is *not* explained by use-dependent changes in outward current" (52). In rat ventricular myocytes flecainide had no effect on the inward rectifier IK_1 (agreement here in all three species) but did block I_{TO} (IC_{50} 3.7 μM) and I_K (IC_{50} 15μM). In contrast to the reported effect in canine tissue, in the rat the effect of flecainide on I_{TO} was frequency dependent, and "the results showed ... that inhibition is mediated through preferential interaction with the open channel" (53).

These results illustrate the problem of extrapolation. Not only may there be species differences, but isolated myocytes, deprived of gap-junction connections with their neighbors and of other external influences and subjected to complex patterns of voltage clamp, may exhibit behavior unlike that of a myocardial cell in a beating human heart.

The enantiomers of flecainide have similar electrophysiological effects on V_{max} in guinea pig papillary muscle (54). Flecainide strongly inhibited the enzyme $P450IID_6$, which could lead to excess sensitivity in some patients (55).

Moricizine

Ethmozine (EN 313, later moricizine) was introduced in the USSR more than a quarter of a century ago, and was shown to be an effective antiarrhythmic drug, especially for supraventricular nodal tachycardia. Moricizine widened QRS, increased both AH and HV intervals (AH>HV), but had little effect on ERP, JT interval, or QT_C (QT corrected for heart rate). On these criteria moricizine would be classed as a 1c compound, but it had been reported to have properties similar to those of 1a and 1b drugs. The significance of onset/offset kinetics in explaining clinical effects had not been appreciated at the time of the early studies on moricizine, but close scrutiny of the Russian data revealed that the onset of its frequency-dependent depression of V_{max} required at least 10 stimuli and that the recovery time was very long (>1 minute) and was independent of concentration, which supported clinical evidence for a 1c classification (56). Moricizine does not lengthen APD, in spite of Russian reports that potassium current was inhibited. In a study on newborn puppies, the onset of frequency-dependent depression of V_{max} in Purkinje fibers was reported to require >23.5 stimuli at a frequency of 5 Hz after a dose of 3 mg.kg^{-1} moricizine, and the recovery-time constant, measured in isolated myocytes, was 10 to 30 s at 16° C (57). CAST 2 with moricizine was terminated before its Class 1c status had been fully accepted.

Cibenzoline

The compound UP 339-01, later named cibenzoline, was introduced by UPSA Laboratories two decades ago, and early unpublished studies established it as a Class 1 antiarrhythmic agent on isolated rabbit atria and as a local anesthetic on desheathed frog nerve equipotent with procaine. Clinical trials in France and elsewhere demonstrated antiarrhythmic efficacy in both supraventricular and ventricular arrhythmias. Its clinical profile (widened QRS, increased HV conduction time, QT prolonged only by the additional QRS width) placed it as a Class 1c compound, confirmed by *in vitro* studies, indicating slow onset/offset kinetics. Cibenzoline had no antiadrenergic action, but possessed about 1/5 the activity of verapamil in antagonizing the positive inotropic effect of raised [Ca]$_o$. In isolated rabbit cardiac preparations cibenzoline significantly prolonged both AH and HV conduction times (58).

Hypoxia shortens APD in atrium and ventricle, even in the presence of glucose and

20% oxygen, which prevent any reduction of intracellular ATP concentration. Cibenzoline greatly *reduced* APD-shortening by successive periods of hypoxia in spontaneously beating isolated rabbit atria. APD-shortening may be responsible, at least as one factor, in the additional risk of arrhythmias associated with ischemic heart disease (IHD), but a trial of cibenzoline in comparison with other antiarrhythmic drugs in IHD has not been undertaken, nor is likely to be in view of the CAST results. In a recent double-blind randomized trial in patients with paroxysmal atrial fibrillation cibenzoline was as effective as flecainide (59).

Lorcainide

Introduced more than a decade ago, lorcainide has been used as an antiarrhythmic drug in Europe, mainly for the treatment of ventricular arrhythmias. *In vitro* studies and its clinical profile established lorcainide as another Class 1c compound, with slow onset (13 s) and offset (16 S) kinetics (60), reported in a comprehensive review of its properties by Janssen. CNS side effects were common.

Propafenone

The structure of propafenone is similar to that of many beta blocking drugs, but with substituents on the nitrogen of the propanolamine (eg, *n*-propyl in propafenone; isopropyl in oxprenolol), and on the ring ($-C:O.CH_2.CH_2$. Phe with propafenone; $-O.CH_2.CH:CH_2$ with oxprenolol, both in the ortho position) which increased the local anesthetic activity of propafenone relative to its beta blocking potency. In various preparations propafenone was estimated to have a beta blocking activity ($pA_2 = 6.3$ to 6.55) about 1/40 to 1/80 that of propranolol ($pA_2 = 8.3$), but since the therapeutic concentration of propafenone is about fifty times greater than that of propranolol, beta blockade must make a substantial contribution to the overall action of propafenone

(61). The beta blocking activity of propafenone, as with propranolol, resides in the S-enantiomer, both enantiomers being equipotent for their local anesthetic effect (62).

By clinical criteria propafenone's profile places it in Class 1c. In long-term administration propafenone increased PR interval 16% to 27%, and QRS duration 18% to 26%. The corrected QT and the JT intervals were not significantly affected. Similar results were recorded by several authors (63)

As a local anesthetic on the guinea pig cornea propafenone was equipotent with procaine. It depressed V_{max} in atrial and ventricular muscle and in Purkinje fibers. The effect was frequency dependent with a slow recovery-time constant equal to that of flecainide (15.5 s), but with a faster onset time constant (4 s) equal to that of disopyramide. Propafenone reduced the spontaneous frequency of the isolated rabbit sinus node, principally by *delaying repolarization*, only the higher concentrations reducing the slope of the slow diastolic depolarization, and APD was prolonged in atrial and ventricular muscle and in Purkinje cells. Finally propafenone reduced the positive inotropic effect of raising $[Ca]_o$ with a potency about 1/20 that of verapamil. Thus propafenone possessed all four major classes of antiarrhythmic action to some extent, by far the most important being the Class 1c action (>2>3>4) (61). Nevertheless the additional actions of propafenone distinguish it quite clearly from the other 1c compounds already described.

The above results on rabbit tissue were substantially confirmed in canine Purkinje fibers, but the effect of propafenone on APD was frequency independent in contrast to the positive frequency dependence with flecainide (64). In voltage-clamp studies on isolated rabbit myocytes it was concluded that propafenone blocked all three major potassium currents, transient outward I_{TO}, inward rectifier I_{K1}, and delayed rectifier I_K. "Propafenone blockade of time-dependent potassium currents is open-state dependent and is relatively selective for I_K" (65).

The effect of propafenone in delaying repolarization by selective block of delayed rectifier potassium channels, coupled with its beta blocking action (with concomitant slowing of SAN frequency and AV conduction delay) account for its efficacy in supraventricular arrhythmias, while the Class 1c action provides protection against ventricular arrhythmias.

MAGNESIUM

An intravenous injection of 1 to 2 g magnesium can sometimes cut short an episode of ventricular tachycardia—for example, during surgery. It was shown by Hanck and Sheets (66) that magnesium enters the sodium channel, impeding access of sodium to the ion pathway. A concentration of 10 mM Mg^{2+} reduced peak sodium current by 46% and shifted the voltage current relation in a positive direction by 20 mV. Magnesium can thus be regarded as a short-term Class 1 agent for immediate use.

SITE OF ATTACHMENT TO SODIUM CHANNELS

The properties of sodium channels have been studied in a variety of preparations, squid axons, frog nerves, snail neurones, isolated single cells, patches of membrane from various tissues, and *Xenopus* oocytes expressing human heart Na^+ channels. Although it is reasonable to assume that all voltage-dependent channels have features in common, an ion-permeable core, a selectivity filter, a charged "voltage sensor," and structures controlling the opening and inactivation of the pathway, it is possible that cardiac channels differ in some respects from those found in other organs and that the effect of a local anesthetic on a nerve is not identical to that of a Class 1 drug on a myocardial cell. For example, although tetrodotoxin (TTX) blocks sodium channels in rabbit Purkinje cells, the concentration required is 1,000 to 10,000 times that sufficient to inhibit conduction in nerve. Local anesthetics were once supposed

to cross nerve membranes in their un-ionized state, but to block sodium channels by entering the ion pathway from the inside and binding to it in their ionized state.

A different mechanism was suggested for sodium channel block by Class 1 drugs in cardiac cells (67). "They have a variety of chemical structures but are mostly lipophilic" and "are taken up into the primary membrane, from where they press their attack on the sodium channels. Class 1 action by the local anesthetic type of drug may involve restriction of the freedom of charged elements controlling the ion gates to move in response to changes in voltage across the membrane." Hille, while still retaining the view that local anesthetics must bind to the interior of the ion pathway to block conduction, suggested that lipophilic compounds entered the membrane and passed from there through the wall of the channel to reach their *attachment site in the pore* (68). More recently Katz et al. (69) demonstrated with ligand probes that various compounds were indeed located within the lipophilic interior of the membrane.

The lipophilic properties of many cardiac drugs appear to play an important role in their interaction with specific receptor sites in the cardiac sarcolemmal membrane. Recent evidence suggests that some of these drugs approach their receptors by a two-step process in which the ligand first partitions into the bulk lipid bilayer of the sarcolemmal membrane, which has a relatively high lipid content. Such drugs can then become precisely oriented in the bilayer, within which they diffuse laterally through the bulk lipids to reach a specific protein receptor site also within the bilayer.

The two mechanisms of action, a physical effect and binding to a specific site, are not mutually exclusive, and it is possible that both are involved in varying proportion in the action of different Class 1 agents. The involvement of both electrostatic and hydrophobic components in the action of antiarrhythmic drugs has recently been studied in human hKv1.5 channels expressed in L cells (70).

Hodgkin and Huxley suggested that sodium channels in the squid giant axon were opened when three charged elements (M) moved independently in response to a positive voltage shift and that inactivation occurred when a single element (H) closed it. A model (1972) of a cardiac sodium channel based on the H-H system, with a central conducting pore and the M and H voltage sensors located inside the wall, is reproduced in Fig. 5. Access of a Class 1 drug is shown as via the lipophilic layer to a site on the wall (67).

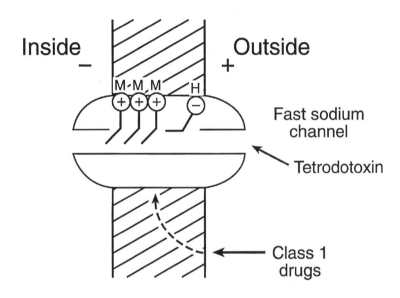

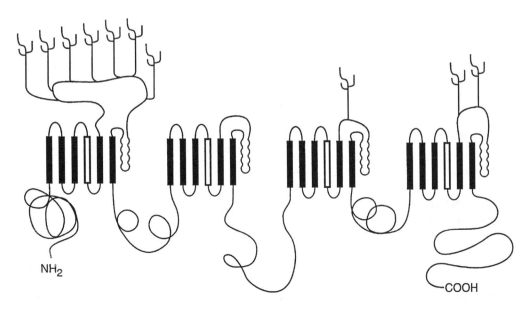

FIG. 5. Models of sodium channels. *Above:* 1972. *Below:* 1992

Since the process of channel opening involves a movement of charge, a "gating current" should be detectable if it could be separated from other charge transfers, as displayed in evidence presented by Armstrong and Benzanilla (71). The amino-acid sequence and tertiary structure of the sodium channel has been elucidated (1), comprising four sets of six components lodged in the membrane and connected by intracellular and extracellular loops. Together with other associated structures, the sets surround an ion channel (Fig. 5). The location of the voltage sensors in the fourth of the six components, and of the inactivation mechanism in one of the intracellular loops (ball-and-chain model) has led to analyses of the molecular details of the operation of the channel beyond the scope of this discussion (72).

The fact that some Class 1 drugs have no effect on sodium channels in their resting state implied that the site of attachment was revealed only after depolarization, an early suggestion amplified into a "guarded receptor" model. Since there are at least four recognized states of the channel, resting (R), activated or open (A), inactivated (I), and secondarily inactivated (SI), and possibly intermediate states as well, more complex models have been constructed, with different drugs possessing specific affinities for each state. If a drug, for example, had no effect either on the resting state or after a positive-clamp pulse sufficiently long to activate sodium current but immediately repolarized, it could reasonably be concluded that its affinity for open channels was low. Various patterns of pulse duration, prepulses, different holding potentials, and so on, in the presence of agents to block unwanted currents, have been employed in an attempt to characterize antiarrhythmic drugs by their affinities for each state. The appropriate modulated receptor model could then be constructed.

In patch-clamp studies positive voltages are applied in response to which a sodium channel may or may not open. If it opens, the channel conductance can be calculated. When it closes, the question arises whether it inevitably proceeds to its inactivated state or could open again without doing so. If it does not open ("blank sweep"), can it pass into its inactivated state without opening? Mitsuiya and Noma concluded that cardiac sodium channels never inactivate without passing through an open state and attributed blank sweeps to failure to observe "late openings" by the use of insufficiently long depolarizations (73). This finding is consistent with the view that plateau sodium current is provided by late openings rather than late inactivations and has implications for the construction of modulated receptor models. The rate constants in such models, for the attachment of drugs to and detachment from the channels in their various states, are not independently determined but selected to fit the final result.

Eventually the precise location of the attachment of Class 1 drugs to sodium channels and their affinities for the various states will doubtless be determined. It will still be necessary, however, to measure what effect the drugs have on the action potentials of the SAN, atrium, AVN, His-Purkinje system, and ventricle; to evaluate their overall influence on human cardiac performance; and to reveal unwanted side-effects. At present, whether a drug attaches to a sodium channel in its open or inactivated state seems less important from the clinical standpoint than how quickly it blocks the channel and for how long.

REFERENCES

1. Catteral WA. Cellular and molecular biology of voltage-gated sodium channels. *Physiol Rev* 1992; 72:515–548.
2. Peters NS, Green CR, Poole-Wilson PA, Severs NJ. Cardiac arrhythmias and the gap junctions. J Mol Cell Cardiol 1995; 27:37–44.
3. Winslow RL, Jongsma HJ. Role of tissue geometry and spatial localization of gap junctions in generation of the pacemaker potential. *J Physiol* 1995; 487:126P.
4. De Mello WC. The cardiac renin-angiotensin system: Its possible role in cell communication and impulse propagation. *Cardiovasc Res* 1995; 29:730–736
5. Sepp R, Severs NJ, Gourdie RG. Altered patterns of cardiac intercellular junction distribution in hypertrophic cardiomyopathy. *Heart*; 1996; 76:412–417.
6. Bigger JT, Mandel WJ. Effect of lidocaine on the electrophysiological properties of ventricular muscle and Purkinje fibers. *J Clin Invest* 1970; 49:63–77.

7. Davis LD, Temte JV. Electrophysiological actions of lidocaine on canine ventricular muscle and Purkinje fibers. *Circ Res* 1969; 24:639–655.

8. Bassett AL, Hoffman BF. Anti-arrhythmic drugs: electrophysiological actions. *Ann Rev Pharmacol* 1971; 11:143–170.

9. Singh BN, Vaughan Williams EM. Effect of altering potassium concentration on the action of lidocaine and diphenylhydantoin on rabbit atrial and ventricular muscle. *Circ Res* 1971; 29:286–295.

10. El-Sherif N, Scherlag BJ, Lazzara R, Hope RR. Reentrant ventricular arrhythmias in the late myocardial infarction period. 4. Mechanism of action of lidocaine. *Circulation* 1977; 56:395–405.

11. Birkhead JS, Vaughan Williams EM. Dual effect of disopyramide on atrial and atrioventricular conduction and refractory periods. *Br Heart J* 1977; 39:657–660.

12. Ronfeld RA. Comparative pharmacokinetics of new antiarrhythmic drugs. *Am Heart J* 1980; 100:978–983

13. Colatsky TJ. Mechanism of action of lidocaine and quinidine on action potential duration in rabbit cardiac Purkinje fibers. *Circ Res* 1982; 50:17–27.

14. Campbell TJ, Vaughan Williams EM. Voltage- and time-dependent depression of maximum rate of depolarization of guinea-pig ventricular action potentials by two new antiarrhythmic drugs, flecainide and lorcainide. *Cardiovasc Res* 1983; 17:251–258.

15. Kidwell GA, Greenspon AJ, Greenberg RM, Volosin KJ. Use-dependent prolongation of ventricular tachycardia cycle length by Type I antiarrhythmic drugs in humans. *Circulation* 1993; 87:118–125.

16. Investigators. Cardiac Arrhythmia Pilot Study (CAPS). *Am J Cardiol* 1986; 57:71–95.

17. Cardiac Arrhythmia Pilot Study Investigators. Effects of encainide, flecainide, imipramine and moricizine on ventricular arrhythmias during the year after acute myocardial infarction: the CAPS. *Am J Cardiol* 1988; 61:501–509

18. Cardiac Arrhythmia Suppression Trial Investigators. Preliminary Report: effect of encainide and flecainide on mortality in a randomized trial of arrhythmia suppression after myocardial infarction. *N Engl J Med* 1989; 321:406–412.

19. Vaughan Williams EM. Classification of the antiarrhythmic actions of moricizine. *J Clin Pharmacol* 1991; 31(3):216–221.

20. Greene HL, Roden DM, Katz RJ, Woosley RL, Salerno DM, Hewthorn RW, and the CAST investigators. The Cardiac Arrhythmia Suppression Trial: First CAST ... Then CAST 2. *J Am Coll Cardiol* 1992; 19:894–898

21. Ruskin JM. The Cardiac Arrhythmia Suppression Trial. *N Engl J Med* 1989; 321:386–388.

22. Bigger JT. Implications of the cardiac arrhythmia suppression trial for antiarrhythmic drug treatment. *Am J Cardiol* 1990; 65:3D–10D.

23. Gottlieb SS. The use of antiarrhythmic agents in heart failure: implications of CAST. *Am Heart J* 1989; 16:1074–1077.

24. Coplen SE, Antman EM, Berlin JA, Hewitt P, Chalmers TC. Efficacy and safety of quinidine therapy for maintenance of sinus rhythm after cardioversion. *Circulation* 1990; 82:1106–1116.

25. Fautrez VM, Adamantidis MM, Caron JF, Libersa CC, Dupuis BA. Comparative electrophysiological effects of metabolites of quinidine and hydroquinidine. *J Cardiovasc Pharmacol* 1992; 19:308–318.

26. Haberman RJ, Rials SJ, Stohler JL, Marinchak RA, Kowey PR. Evidence for a re-excitability gap in man after treatment with Type I antiarrhythmic drugs. *Am Heart J* 1993; 126:1121–1126.

27. Rosenheck S, Sousa J, Calkins H, Kadish AH, Morady F. The effect of rate on prolongation of ventricular refractoriness by quinidine in humans. *Pace* 1990; 13: 1379–1384.

28. Nademanee K, Stevenson WG, Weiss JN, Frame VB, Antimisiaris MG, Suithichakul T, Pruitt CM. Frequency dependent effects of quinidine on the ventricular action potential and QRS duration in humans. *Circulation* 1990; 81:780–796.

29. Leitch JW, Klein GJ, Yee R, Feldman RD, Brown J. Differential effect of intravenous procainamide on anterograde and retrograde accessory pathway conduction. *J Am Coll Cardiol* 1992; 19:118–124.

30. Villemaire C, Savard P, Talajic M, Nattel S. A quantitative analysis of use-dependent conduction slowing by procainamide in anesthetized dogs. *Circulation* 1992; 85:2255–2266.

31. Lee RJ, Liem LG, Cohen TJ, Franz MR. Relation between repolarization and refractoriness in the human ventricle: cycle length dependence and effect of procainamide. *J Am Coll Cardiol* 1992; 19:614–618.

32. Buxton AE, Josephson ME, Marchlinski FE, Miller JM. Polymorphic ventricular tachycardia induced by programmed stimulation: response to procainamide. *J Am Coll Cardiol* 1993; 21:90–98.

33. Kulakowski P, Bashir Y, Heald S, Paul V, Anderson MH, Gibson S, Malik M, Camm AJ. Effects of procainamide on the signal-averaged electrocardiogram in relation to the results of programmed ventricular stimulation in patients with sustained monomorphic ventricular tachycardia. *J Am Coll Cardiol* 1993; 21:1428–1438.

34. Goldstein S, Brooks MM, Ledingham R, Kennedy HL, Epstein AE, Pawitan Y, Bigger JT. Association between ease of suppression of ventricular arrhythmia and survival. *Circulation* 1995; 91:79–83.

35. Sekiya A, Vaughan Williams EM. A comparison of the antifibrillatory actions and effects on intracellular cardiac potentials of pronethalol, disopyramide and quinidine. *Br J Pharmacol* 1963; 21:473–481.

36. Vaughan Williams EM. Disopyramide. *Ann N Y Acad Sci* 1984; 432:189–200

37. Birkhead JS, Vaughan Williams EM. Dual effect of disopyramide on atrial and atrioventricular conduction and refractory periods. *Br Heart J* 1977; 33:657–660.

38. Vaughan Williams EM. A classification of antiarrhythmic actions after a decade of new drugs. *J Clin Pharmacol* 1984; 24:129–147.

39. Ju Y-K, Saint DA, Gage PW. Effects of lignocaine and quinidine on the persistent sodium current in rat ventricular myocytes. *Br J Pharmacol* 1992; 107: 311–316.

40. Furukawa T, Koumi S-L, Sakakbara Y, Singer DH, Jia H, Arentzen CE, Backer CL, Wasserstrom JA. An analysis of lidocaine block of sodium current in isolated human atrial and ventricular myocytes. *J Mol Cell Cardiol* 1995; 27:831–846.

41. Aupetit JF, Timour Q, Loufoua-Moundanga J, Barral-Cadiere L, Lopez M, Freys ZM, Faucon G. Profibrillatory effects of lidocaine in the acutely ischemic porcine heart. *J Cardiovasc Pharmacol* 1995; 25:810–816.

42. Singh BN, Vaughan Williams EM. Investigations of the mode of action of the new antidysrhythmic drug, Kö 1173. *Br J Pharmacol* 1972; 44:1–9.
43. Sato T, Shigematsu S, Arita M. Mexiletine-induced shortening of the action potential duration of ventricular muscles by activation of ATP-sensitive K^+ channels. *Br J Pharmacol* 1995; 115:381–382.
44. Sato T, Shigematsu G, Arita M. Mexiletine activates the ATP-sensitive K^+ channel and protects the heart from myocardial stunning. *J Mol Cell Cardiol* 1995; 27: A517(64).
45. Sunami A, Fan Z, Sawanabori T, Hiraoka M. Use-dependent block of Na^+ currents by mexiletine at the single channel level in guinea-pig ventricular myocytes. *Br J Pharmacol* 1993; 110:183–192.
46. Cowan JC, Vaughan Williams EM. Characterization of a new antiarrhythmic drug, flecainide (R818). *Eur J Pharmacol* 1981; 73:333–342.
47. Katritsis D, Rowland E, O Nunain S, Shakespeare LF, Poloniecki J, Camm AJ. Effect of flecainide on atrial and ventricular refractoriness and conduction in patients with normal left ventricle. *Eur Heart J* 1995; 16: 1930–1935.
48. Chimient M, Cullen MT, Casadei G. Safety of flecainide versus propafenone for long-term management of symptomatic paroxysmal supraventricular tachyarrhythmias. *Eur Heart J* 1995; 16:1943–1951.
49. Goldberger J, Helmy I, Katzung B, Scheinman M. Use-dependent properties of flecainide acetate in accessory atrioventricular pathways. *Am J Cardiol* 1994; 73: 43–49.
50. Wang Z, Pelletier LC, Talajic M, Nattel S. Effects of flecainide and quinidine on human atrial action potentials. *Circulation* 1990; 82:274–283.
51. Follmer CH, Colatsky TJ. Block of delayed rectifier potassium current, I_K, by flecainide and E-4031 in cat ventricular myocytes. *Circulation* 1990; 82:289–293.
52. Wang Z, Fermini B, Nattel S. Mechanism of flecainide's rate-dependent actions on action potential duration in canine atrial tissue. *J Pharmacol Exp Ther* 1993; 267:575–581
53. Slawsky MT, Castle NA. K^+ channel blocking actions of flecainide compared with those of propafenone and quinidine in adult rat ventricular myocytes. *J Pharmacol Exp Ther* 1994; 269:66–74.
54. Vanhoutte F, Vereecke J, Carmeliet E, Verbeke N. Effects of enantiomers of flecainide on action potential characteristics in the guinea-pig papillary muscle. *Arch Int Pharmacodyn* 1991; 310:102–115.
55. Haefeli WE, Bargetzi MJ, Follath F, Meyer UA. Potent inhibition of cytochrome P450IID₆ (debrisoquine 4-hydroxylase) by flecainide in vitro and in vivo. *J Cardiovasc Pharmacol* 1990; 15:776–779.
56. Vaughan Williams EM. Classification of the antiarrhythmic actions of moricizine. *J Clin Pharmacol* 1991; 31:216–221.
57. Liu W, Clarkson CW, Yamasaki S, Chang C, Stolfi A, Pickoff AS. Characterization of the rate-dependent effects of ethmozine on conduction, in vivo, and on the sodium current, in vitro, in the newborn heart. *J Pharmacol Exp Ther* 1992; 263:608–616.
58. Millar JS, Vaughan Williams EM. Effects on rabbit nodal, ventricular and Purkinje cell potentials of a new antiarrhythmic drug, cibenzoline, which protects against action potential shortening in hypoxia. *Br J Pharmacol* 1982; 75:469–478.
59. Babuty D, D'Hautefeuille B, Scheck F, Mycinsky C, Pruvost P, Perauderu P. Cibenzoline versus flecainide in the prevention of paroxysmal atrial arrhythmias: a double-blind randomized study. *J Clin Pharmacol* 1995; 35:471–477.
60. Vaughan Williams EM. The relevance of the classification of antiarrhythmic drugs to their clinical use: the place of lorcainide. In: *Lorcainide*. Brussels: Janssen; 1986:5–15.
61. Dukes ID, Vaughan Williams EM. The multiple modes of action of propafenone. *Eur Heart J* 1984; 5:115–125.
62. Stochitzky K, Klein W, Stark G, Stark U, Zernig G, Graziedei I, Lindner W. Different stereoselective effects of (R)- and (S)-propafenone: clinical, pharmacologic, electrophysiologic and radioligand binding studies. *Clin Pharmacol Ther* 1990; 47:740–746.
63. Vaughan Williams EM. Significance of classifying antiarrhythmic actions since the Cardiac Arrhythmia Suppression Trial. *J Clin Pharmacol* 1991; 31:123–135.
64. Zaza A, Malfatto G, Schwartz PJ. Diverse electrophysiologic effects of propafenone and flecainide in canine Purkinje fibers: implications for antiarrhythmic drug classification. *J Pharmacol Exp Ther* 1994; 269: 336–343.
65. Fermini DDB, Nattel S. Potassium channel blocking properties of propafenone in rabbit atrial myocytes. *J Pharmacol Exp Ther* 1992; 264:1113–1123.
66. Hanck DA, Sheets MF. Extracellular divalent and trivalent cation effects on sodium current kinetics in single canine cardiac Purkinje cells. *J Physiol* 1992; 454:267–298.
67. Vaughan Williams EM. Biophysical background to beta-blockade. In: *New Perspectives in Beta-blockade*. Ed., DM Burley. Horsham, England:CIBA; 1972:11–39.
68. Hille B. Local anaesthetics: hydrophilic and hydrophobic pathways for the drug-receptor reaction. *J Gen Physiol* 1977; 69:487–515.
69. Katz AM, Rhodes DG, Herbette LG. Role of membrane bilayer in ligand-receptor interactions. *J Mol Cell Cardiol* 1986; (Suppl 1) 18:33a (abst).
70. Snyders DJ, Yeola SW. Determinants of antiarrhythmic drug action. *Circ Res* 1995; 77:575–583.
71. Armstrong CM, Benzanilla F. Currents related to movement of the gating particles of the sodium channels. *Nature* 1973; 242:459–461.
72. Yang N, George AL, Horn R. Molecular basis of charge movement in voltage-gated sodium channels. *Neuron* 1996; 16:113–122.
73. Mitsuiya T, Noma A. Inactivation of the cardiac Na^+ channels in guinea-pig ventricular cells through the open state. *J Physiol* 1995; 485:581–594.

3

Class 2 Antiarrhythmic Action

E. M. Vaughan Williams

AUTONOMIC CONTROL OF HEART RATE

The resting oxygen requirement of the body is a function of age and surface area. The basal metabolic rate (BMR) declines through life, as does the cardiac output required to supply it, a normal cardiac index falling from 4.2 $l.min^{-1}.m^{-2}$ at five to ten years to 2.5 $l.min^{-1}.m^{-2}$ at sixty. If the resting cardiac output (CO) were 6.0 $l.min^{-1}$ and the stroke volume (SV) were 100 ml, heart rate (HR) would be 60 beats.min^{-1}. At rates greater than 180 to 200 beats.min^{-1} CO reaches a limit, because a higher rate would leave insufficient time for the ventricles to fill and empty at the same SV (Stroke Volume). Trained athletes have hypertrophied hearts, which may take two forms: eccentric, with a large left ventricular mass and chamber produced by exercise with a high dynamic component (track running); or concentric, produced by static exercise (weight lifting). Intermediate forms occur in cyclists and rowers. At rest, the hypertrophied heart has a larger SV, but the body's oxygen requirement is little increased, so that an adequate cardiac output can be achieved at a lower HR. At peak exercise, therefore, CO can be augmented four-fold or more at the maximum HR, with oxygen uptake increasing in parallel. In a group of elite cyclists, for example, maximum uptake (VO_2) was 66.2 ($\pm$ 4.4) $ml.kg^{-1}.min^{-1}$, compared with 49.1 ($\pm$7.3) $ml.kg^{-1}.min^{-1}$ in controls (1).

Cholinergic Influence

In the absence of sinoatrial disease the resting HR is an indicator of physical fitness and the sinus node is predominantly under vagal control with little sympathetic involvement until HR increases on exercise to the point of withdrawal of vagal influence. The mechanism by which acetylcholine (Ach) released from the vagus nerve reduces HR was first demonstrated by Hutter and Trautwein in the frog sinus venosus (2), which was hyperpolarized, preventing or delaying the development of the slow diastolic depolarization. The hyperpolarization is now known to be due to the opening of potassium-selective channels, through which flows an outward current, I_{K-Ach}. These channels are present in the SAN (Sino Atrial Node), AVN (Atrial Ventricular Node), and atrial myocardium of all mammalian species studied, but absent from cardiac Purkinje cells. Activation of Ach receptors in the atrium depresses contractions, but the effect of Ach on the ventricle has been controversial. Ach receptors occur on postganglionic sympathetic nerves, and their stimulation reduces noradrenaline release, so that some reports of a negative inotropic effect on the ventricle may have been attributable to depression of sympathetic activity. In the canine ventricle it was observed that application of Ach to *endocardial* muscle had no effect, but it shortened APD (Action Potential Duration) and depressed contractions in the *epicardium* (3). Epicardial action potentials have a "spike and wave" form, the peak of depolarization being followed by a brief repolarizing dip before the development of the plateau. A similar notch occurs in the action potential of proximal Purkinje cells. Various explanations have been given since 1961 for the appearance of this transient current (discussed in the next chapter), but there is now consensus that the charge carrier is mainly potassium, providing a "transient outward current," I_{TO}, the function of which is unclear. The Ach-induced repolarization and negative inotropic effect in canine epicardium were not due to activation of I_{TO}, however, because they were unaffected by 4-aminopyridine (4-AP), which blocks I_{TO}.

In the atrium of the bull frog the negative inotropic action of Ach was reported to be caused by block of inward calcium current. The second, or slow inward current, isi, was found to be markedly reduced by concentrations of acetylcholine greater than approximately 2×10^{-8} M. The resulting decrease in net calcium entry provides a straightforward explanation for the negative inotropic action of acetylcholine in atrial muscle (4). Cholinergic action in reducing HR in the frog sinus venosus was likewise attributed to inhibition of isi (5). In canine epicardial cells, however, even very high concentrations of Ach (10^{-5} M) reduced inward calcium current by only 8% (3). In isolated ferret ventricular cells, which are sensitive to Ach, a negative inotropic action identical to that induced by Ach was achieved in a drug-free medium by clamping the cells to receive a wave-form of depolarization exactly matching the shortened APD observed in the presence of Ach. It was concluded that the negative inotropic action of Ach was caused by the APD shortening, which cut off the activation of ICa, and that there was no direct inhibition of calcium current (6).

Delayed conduction through the AV node in response to vagal stimulation can likewise be attributed to cut-off of inward calcium current by the shortened APD, because depolarization in the AVN is by calcium, not sodium current. Adenosine also blocks AV conduction by opening potassium channels and is an effective antiarrhythmic in supraventricular arrhythmias, but has a very short duration of action. Ach is rapidly removed by high local concentrations of specific cholinesterase (AchE), and any transmitter that escapes is mopped up by a less specific plasma cholinesterase.

Adrenergic Influence

In contrast to the simplicity of parasympathetic action, sympathetic control is much more complex. Action potentials descending the postganglionic axons promote calcium entry, which activates vesicles situated in nerve swellings ("varicosities"). The vesicles stick to the inside of the membrane forming omega structures from which granules are released (exocytosis) into the extracellular space, exuding noradrenaline, ATP, and neuropeptide Y, together with other components without transmitter functions. At some sites, such as the vas deferens of the guinea pig, the ATP, via a purine receptor, induces an excitatory postsynaptic potential. ATP release can be inhibited by prostaglandins E_1 or E_2 (7). "Overflow of catecholamines and purines from chromaffin cells occurred in a constant ratio of approximately 4 to 1." The ratio of catecholamine to purine release from sympathetic varicosities varied according to the pattern of stimulation, leading to the conclusion that "the two transmitters are released from two separate populations of exocytotic vesicles" (8).

The function of neuropeptide Y (NPY) is uncertain, some authors reporting that it facilitates adrenergic transmission by activation of inward calcium current, but Bryant and Hart found that very low concentrations of neuropeptide Y (10^{-6} M) *reduced* I_{Ca} in guinea pig ventricular myocytes, via a G-protein which inhibited the formation of cyclic adenosine monophosphate (cAMP) (9). Neuropeptide Y (NPY) had no direct effect on normal sinus rhythm in guinea pig atria, nor did it alter the action of noradrenaline. "A prejunctional action to inhibit release of acetylcholine from parasympathetic nerve endings is implied" (10).

Noradrenaline acts on several types of receptor. It is removed much less rapidly than Ach, so that strong sympathetic stimulation causes transmitter to overflow into the plasma. Noradrenaline is taken up again by the nerve to be reused or metabolized by monoamine oxidase (MAO). It can also enter the postsynaptic cell, to be removed by catechol-o-methyltransferase (COMT). Some of the released noradrenaline exerts negative feedback to the release of further transmitter by stimulation of presynaptic α_2-adrenoceptors. There is also parasympathetic inhibition of adrenergic stimulation via muscarinic receptors on sympathetic nerves. Finally, an interaction exists between the sympathetic and renin-angiotensin-aldosterone systems, which is not yet fully defined. Embryonic heart muscle incubated in an eye develops angiotensin receptors (A_1) if the eye is sympathetically denervated, but not so many if sympathetic supply is intact, implying negative feedback on the angiotensin system by the sympathetic system (11). Conversely, angiotensin potentiates some sympathetic effects, so that angiotensin converting enzyme (ACE) inhibitors may exert indirectly a Class 2 antiarrhythmic action (12).

There are three types of β-adrenoceptor, of which β_1 and β_2 are the most important in the heart, β_3 receptors mediating effects mainly at extracardiac sites. The ratio of the density of $\beta_1 : \beta_2$ adrenoceptors is 70% : 30% in atria, and 80% : 20% in ventricles, the β_1 density being 55 fM.mg^{-1} protein in both chambers. β_2 density was 25 fM.mg^{-1} in atria, and 15 fM.mg^{-1} in ventricle. Receptors are continuously being synthesized and reabsorbed, so that the density may be increased ("up-regulation") or decreased in response to various conditions (13). "In patients with severe heart failure due to different causes, including idiopathic dilated cardiomyopathy and ischemic myopathy, the total β-adrenoceptor density is reduced. In patients with end-stage idiopathic dilated cardiomyopathy the β_1-adrenoceptor population is selectively reduced."

The question arises whether β_3-adrenoceptors are functionally relevant in cardiac muscle. The compound (–)-CGP 12177 (CGP) is an antagonist of β_1- and β_2-adrenoceptors, but is an agonist at β_3-receptors. In a human atrial appendage obtained during surgery, in the presence of 200 nM of the β_1- and β_2-receptor blocker propranolol, CGP increased the force of contraction. The maximal effect of CGP was equivalent to 15% of the effect of 400 µM

of isoprenaline, and of 11% of the effect of 6.75 mM $CaCl_2$ (14). In the rat, the cardiac β_3-adrenoceptors were not, however, the same as the β_3-adrenoceptors of the rat colon (15)

Stimulation of β_1 or β_2 adrenoceptors induces a rise in intracellular cAMP concentration, but the fact that the effects of β_1 and β_2 activation are different implies that intracellular compartments or localized structures exist to guide the cAMP to the appropriate destination. The ionic currents involved in SAN pacemaking have been the topic of much controversy, especially with respect to the role of chloride current and its adrenergic activation (discussed in Chapter 6). In the SAN, clusters of small pale cells (p-cells), with very few interconnecting gap junctions and with low maximum diastolic potentials ($\leq$,-50mV), are closely coupled to surrounding transitional cells, which in turn are coupled to more distal atrial myocardial cells.

Selective β_1-adrenoceptor activation has three main effects in the rabbit SAN. 1) APD is shortened (accounting for about 1/3 of the reduction in cycle length), due to opening of potassium channels; 2) the slope of the slow diastolic depolarization is increased (causing further reduction in cycle length); 3) the rate of rise of the upstroke of the depolarization limb is increased. In the SAN calcium ions carry the depolarizing current. In the atrial myocardium contractile force is greatly increased by β_1-receptor stimulation and the rate of development of force is faster, both consistent with an increase of inward calcium current. In contrast, β_2-adrenoceptor stimulation has little positive inotropic action (16). The differential effects of β_1- and β_2-receptor stimulation have recently been confirmed in guinea pig hearts (17). A selective β_1-adrenoceptor agonist (T-0509) at 10^{-8}M concentration increased the rate of development of force (dF/dt max) by 170%, but procaterol, a β_2-receptor selective agonist, at ten times the concentration, increased dF/dt max by only 15%.

Gadsby showed that "β-adrenoceptor agonists" increased K^+ conductance in cardiac Purkinje fibers (18). *Selective β_2-receptor stimulation in the SAN accelerated repolariza-*

tion, but by a different mechanism from that of β_1-receptor stimulation. APD was shortened by stimulation of the Na/K pump, providing net outward current (3 Na^+ out; 2 K^+ in). β_2-receptor stimulation increased the slope of the slow diastolic depolarization, but not of the upstroke of the action potential. The pump is stimulated in skeletal muscle also, and vigorous sympathetic activity causes a fall in plasma potassium.

In contrast to β-receptor stimulation, selective activation of α_1-adrenoceptors *delayed* repolarization and had a small positive inotropic action associated with a slower rate of development of force and a longer lasting contraction (16). α_1-adrenoceptor-mediated effects do not involve the formation of cAMP, and may, therefore, persist in conditions in which β-adrenoceptor stimulation fails. Selective α_2-adrenoceptor stimulation had no effect in any of the cardiac tissues studied (SAN, atrium, AVN, Purkinje cells, ventricle), as is consistent with absence of α_2-receptor binding sites in the myocardium. α_2-adrenoceptors mediate vasoconstriction in small arterioles, α_1-adrenoceptors in larger vessels.

Adrenergic Arrhythmogenicity

An association between stress or emotion and a "heart attack" has long been recognized. A reasonable explanation would be that a sudden increase in blood pressure or pulse pressure in a coronary artery already containing an atheromatous plaque could cause the latter to split and expose the underlying collagen to platelets, which in turn would initiate thrombosis. Adrenergic stimulation can also precipitate cardiac arrhythmias. ECG records obtained during stressful activities such as motor-racing, ski-jumping, even giving a lecture, exhibit extrasystoles and brief runs of ventricular tachycardia (VT). In healthy hearts such abnormalities are self-correcting; but if a susceptible "substrate" exists, the stress may trigger a life-threatening VT or fibrillation (VF), and there are many anecdotes of sudden death in response to an emotional shock or to intense fear.

Background sympathetic activity follows a circadian rhythm, falling to a minimum during sleep, then rising again about 2 hours *before* waking and rising to a peak during the next few hours. The majority of out-of-hospital sudden deaths occur during the morning. Objective evidence of a circadian influence on the incidence of arrhythmias in patients at risk of VT fitted with implantable defibrillators was obtained by recording the time at which the instruments were activated. Forty-two percent of shocks were delivered between 6 and 14 hours (midnight, 0), the highest incidence being at 7 to 12 hours. Only 11% of shocks occurred during the low sympathetic activity period 0 to 6 hours (19). The dependence of arrhythmias on circadian fluctuations of sympathetic activity is illustrated by an 88% predictability for efficacy of beta-blockade in suppressing ventricular ectopic beats (VEB) in patients in whom the incidence of VEBs declined during sleep. If the VEBs persisted during sleep, β-blockers were ineffective, indicating that in these patients the VEBs were not adrenergically driven (20).

There is a circadian rhythm for body temperature also, which falls to a minimum at 6 hours. The major controlling factor that sets the phase of the circadian rhythm is not sleep, activity, or social contact, but light. Experiments by Duffy et al. on human volunteers supported "the concept that the light-dark cycle is the most important synchronizer of the human circadian system. Inversion of the sleep-wake, rest-activity, and social-contact cycles provides relatively minimal drive for resetting the human circadian pacemaker" (21). The circadian response to light is an evolutionary adaptation. Nocturnal animals have an inverse relation to light. Rabbits, which forage at dusk, kept in an environment of 12 hours of light and 12 hours of darkness, exhibit a minimum heart rate 3 to 6 hours after the light is switched on; HR begins to rise 4 hours before the light goes off, peaking 4 hours into darkness (22). The nocturnal Tasmanian Devil also exhibits an inverse circadian rhythm.

There are several mechanisms by which adrenergic activity could precipitate or exacerbate an arrhythmia. An ectopic pacemaker could be activated and induce an out-of-phase depolarization. Acceleration of repolarization shortens APD and refractory period. Some of the extracardiac effects of sympathetic activity increase the probability of arrhythmia. The primary function of β_2-adrenoceptors is to relax the bronchi and peripheral arterioles, but they also stimulate the Na/K pump throughout the skeletal musculature as well as in the heart. In a nonexercising subject noradrenaline induces a fall in plasma K, which is blocked by nonselective β-blockers (blocking β_1- and β_2-adrenoceptors), but not by β_1-selective blockers (23). Low plasma K facilitates arrhythmias, especially in the presence of other arrhythmogenic factors, cardiac glycosides, halogen anesthetics, or ischemia. The additional O_2 consumption provoked by the positive inotropic action of sympathetic stimulation will itself exacerbate ischemia.

Although sympathetic activity is part of normal physiological responses to stress, when it is excessive or inappropriate (eg, in cardiac failure) it can be arrhythmogenic, and drugs that reduce adrenergic stimulation are antiarrhythmic in these circumstances. A definition of *excessive* is arbitrary, but a recent report of an increase in sudden deaths among young Swedish orienteers provides an example. There were sixteen fatalities (15 men, 1 woman) in a group of apparently very fit young people. "No coronary artery disease or anomalies were reported in any of the 16 cases. All performed in their sport at, or close to, their maximum shortly before they died." *Post mortem* examination revealed some myocarditis. The training regime had been vigorous and continuous and was subsequently modified to provide adequate rest periods. "After late 1992 there have been no new unexpected cardiac death cases among young orienteers" (24). The implication is that extreme exertion in the presence of mild myocarditis, such as might be associated with a prior infection, can precipitate fatal arrhythmias in apparently fit youths.

EFFECTS OF INDIVIDUAL CLASS 2 ANTIARRHYTHMIC AGENTS

Sympathetic activity can be reduced in many different ways. The first possibility is to restrict the outflow of signals from the brain to preganglionic neurones. This is achieved by methyldopa and clonidine, but the nerves affected are mainly vasoconstrictor so these drugs are antihypertensives. Second, activity can be blocked at sympathetic ganglia, as by hexamethonium or mecamylamine, but there is no cardiac selectivity. Third, sympathetic endings can be disorganized or destroyed, as by 6-hydroxydopamine, reserpine, or guanethidine. Although the latter two drugs were used for the treatment of hypertension and guanethidine was shown to have antiarrhythmic properties in man, their current use is as pharmacological tools with which to eliminate a sympathetic component ("chemical sympathectomy") in the analysis of the mode of drug actions. Fourth, restriction of release of noradrenaline from sympathetic nerve endings without causing degeneration of the nerves forms part of the antiarrhythmic action of bretylium and amiodarone.

Bretylium

The action of this drug is complex. At first it releases noradrenaline from sympathetic varicosities, causing sinus tachycardia and hypertension. Eventually noradrenaline release in response to postganglionic nerve stimulation is greatly reduced. Bretylium has a negligible local anesthetic action (1/90 that of procaine) or Class 1 action in atrium and ventricle and is not negatively inotropic. In the atrium it does not prolong APD (25), but lengthens ventricular APD. Bretylium, after an initial proarrhythmic action, thus has a dual antiarrhythmic effect, Class 2 and Class 3. Although the drug is still used as a reserve antiarrhythmic, its peripheral antisympathetic action may lead to severe hypotension.

COMPETITIVE ADRENOCEPTOR BLOCKADE

Additional Local Anesthetic Action

The first β-adrenoceptor antagonist, pronethalol, was a derivative of isoprenaline, a naphthaline being substituted for the catechol (Fig. 1). Pronethalol was withdrawn after the discovery of thymic tumors in mice, but not before its antiarrhythmic action had been demonstrated in animals (Chapter 1, Ref. 12) and man. Its successor, propranolol, was a propanolamine. A large number of β-blockers were soon synthesized, some following the ethanolamine structure of pronethalol (Fig. 1, *left*), others the propanolamine of propranolol (Fig. 1, *right*). Propranolol had a local anesthetic activity four times greater than procaine; alprenolol and oxprenolol also had comparable local anesthetic potency.

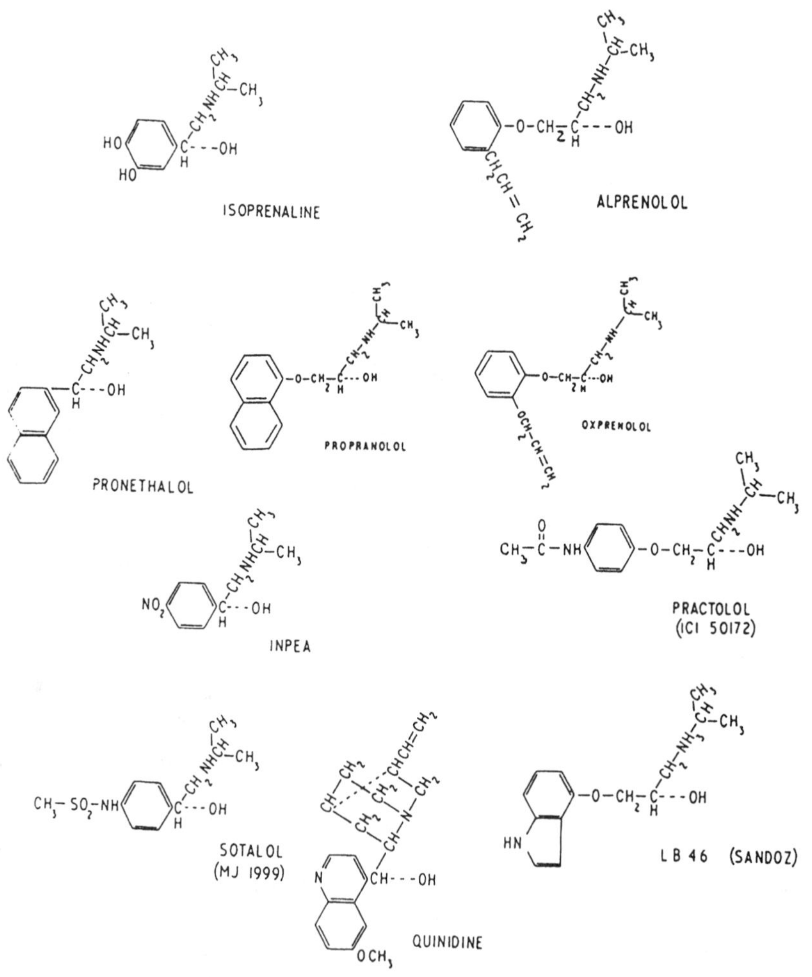

FIG. 1. Structure of isoprenaline and the early β-blockers. *Left,* ethanolamines. *Right,* propanolamines. β-blocking activity resides in the (-)-isomers. Though not a β-blocker, quinidine is more potent than its stereoisomer, quinine. (LB46 = pindolol).

β1-Adrenoceptor Selectivity

Practolol was introduced as an alternative to propranolol, because it had a negligible local anesthetic action at clinical concentrations. It was found subsequently that its effects on cardiac β-receptors greatly exceeded its activity in the bronchi and peripheral vasculature, and it was this discovery which led to the subdivision of β-receptors into β_1 and β_2. Alprenolol and oxprenolol were nonselective. The side chain on both of the latter is attached at the ortho position of the ring, whereas practolol's side chain is at the para position. Structure action studies showed that if the side chain of oxprenolol was transferred from the 2- to the 4-position, the drug became β_1-selective. In fact, though it was never developed as a therapeutic agent, para-oxprenolol was at that time the most β_1-selective compound yet studied (26).

Practolol also had to be withdrawn, due to toxicity in some patients. The side chain of practolol is $CH_3.C:O.NH-$. If the acetic acid is hydrolyzed off, an aniline derivative is left. In atenolol, the successor to practolol, the nitrogen and acetic acid are reversed, the side chain becoming $NH_2.C:O.CH_2-$. Like practolol, atenolol is β_1-selective and lacks local anesthetic activity.

Intrinsic Sympathomimetic Activity (ISA)

Some of the new β-blockers were partial agonists, including oxprenolol and pindolol. The effect was promoted as advantageous. With dubious logic it was claimed that intrinsic sympathomimetic activity (ISA) prevented beta-blockade from "going too far," a steady background sympathetic support being supplied by the drug. If the desired effect is beta-blockade, an excessive response simply requires a reduction of drug concentration, permitting the natural transmitter to restore sympathetic support. In several trials ISA has been found to be deleterious. An early study of long-term treatment with pindolol in post-MI (Myocardial Infarction) patients showed no benefit (27), and a recent review of a variety of procedures to provide "cardioprotection" concluded "an essential principle is the protection of the heart against sympathetic overstimulation and its deleterious consequences. There was no improvement, however, when β-adrenoceptor blockers with intrinsic sympathomimetic activity were used—for example, in the European infarct study with oxprenolol. The cardioprotective effect of β-blockade is directly related to reduction in heart rate" (28).

LONG-TERM β-BLOCKADE AND RATE CORRECTION OF QT

There have been many long-term trials of the effect of β-blockers in post-MI patients, which have consistently demonstrated a reduction in mortality of 25%–30%, but it is not proven that the benefit is solely due to reduction in heart rate and/or a fall in blood pressure. In animals that have been sympathetically denervated, APD in atria and ventricles is prolonged and revealed as a lengthening of QT interval. A similar result is observed after long-term β-blockade in rabbits (29,30) and in man (31). Pacing the heart at increasing frequency shortens the QT interval, and a correction is often made (QT_C) in accordance with an equation dating from 1920 (Bazett's formula), $QT_C = QT/\sqrt{R–R}$ interval. In normal physiological responses to exercise, however, heart rate increases in response to sympathetic stimulation, and since β-receptor stimulation itself shortens APD, such a correction is inappropriate. After long-term β-blockade the slope of the QT/R-R relation becomes less steep (31). A similar flattening of the QT/R-R relation occurs in primary autonomic failure (32). If a comparison of the effect of any drug or procedure on QT is to be valid, it is necessary always to measure QT at the same heart rate, because the appropriate correction for rate cannot be made if the slope of QT/R-R relation varies. This can quite easily be done by exercising a subject until HR exceeds, say, 110 beats.min^{-1}, and then recording the ECG continuously while the heart rate falls subsequently. It is then possible to select a heart rate, say, 100 beats.min^{-1}, at which all measurements of QT interval are made (33).

These findings are relevant to the protection provided by β-blockers in post-MI patients. Infarction kills not only myocardial cells, but also autonomic nerves, as has been demonstrated in dogs after coronary ligation (34). The surviving muscle, beyond the infarcted zone, will have been sympathetically denervated and will have a longer APD than its innervated neighbors, a disparity that is arrhythmogenic. Long-term beta-blockade will prolong APD in the innervated area also, thus restoring homogeneity of APD and removing the source of arrhythmia (35). Another effect of long-term β-blockade in rabbits was to increase capillary density in the ventricle (36,37), the growth perhaps being stimulated by the longer period of coronary blood flow in diastole. Since the technique requires the infusion of fixation fluid into the coronary arteries of beating hearts, the effect cannot be demonstrated in humans.

There are now so many β-blockers that the choice of a suitable agent depends largely upon attendant subsidiary properties. Elimination half-life varies from 20 hours with nadolol, which is nonselective and without local anesthetic activity, to 8 minutes with esmolol, which is also without local anesthetic activity but is β_1-selective. A long half-life facilitates compliance, but increases danger of overdosage, especially in asthmatics.

The major use of β-blockers is as antianginal or antihypertensive agents, and local anesthetic activity is irrelevant. A Class 1 action could, however, add to the antiarrhythmic benefits of β-blockade. It has sometimes been assumed that there is a correlation between the local anesthetic and negative inotropic effects of β-blockers, but this is not supported by a comparison of β-blocking and local anesthetic potencies relative to negative inotropic potency (Table 1).

Propranolol, alprenolol, and oxprenolol are strong local anesthetics, several times more potent than procaine, yet the ratio (Class 1 action on V_{max}) : (negative inotropic action) is actually better than for the Class 1 drugs without β-blocking activity.

BETA-BLOCKING DRUGS WITH ADDITIONAL PROPERTIES

Sotalol

Sotalol (MJ 1999) has a side chain in the para position, but it is not β_1-selective because it is an ethanolamine and so fits the β-receptor differently from the propanolamines. The sulfonamide side chain conferred an additional property, prolongation of APD (38). The β-blocking activity resides in the (–)-isomer, but the effect on APD is exerted by both isomers, which led to the development of the (+)-isomer as a "pure" Class 3 antiarrhythmic drug. Alprenolol is nonselective and has a strong local anesthetic action (39), whereas Pindolol is not a local anesthetic (40), but is notable for its ISA, rendering it unsuitable for long-term prophylactic therapy in post-MI patients.

Labetolol

Labetolol blocks both β- and α-adrenoceptors, a dual action that in theory should make

TABLE 1. *Ratio of β-blocking : Negative inotropic potency and ratio of Class 1 action : Negative inotropic action of various antiarrhythmic agents.*

Order Drug	β-blockade Neg. Inotrop	Order Drug	Class1 Action Neg. Inotrop
1 Pindolol	34,590	1 Sotalol	12.42
2 Sotalol	11,090	2 Oxprenolol	7.05
3 Oxprenolol	2,630	3 Propranolol	6.76
4 (–)Propranolol	2,630	4 INPEA	6.26
5 Practolol	1,260	5 Alprenolol	5.5
6 Alprenolol	1,017	6 Pindolol	4.83
7 INPEA	501	7 Quinidine	3.4
8 (±)Propranolol	21	8 Lidocaine	2.57
		9 Procainamide	2.38
		10 Practolol	1.34

it more effective as an antihypertensive agent. In practice this has not proved to be so, perhaps because of corrective reflex responses. In rabbit hearts it depressed Vmax in atrial and ventricular muscle and in Purkinje cells. It had twice the potency of procaine as a local anesthetic in desheathed frog nerves. Labetolol abbreviated the action potential plateau in normoxic atrial muscle, but attenuated APD shortening induced by hypoxia (cf cinbenzoline). In normoxic ventricular muscle it caused a significant slowing of all phases of repolarization. There was no negative inotropic action in normoxia or hypoxia, and there was no evidence of slowing of A-V conduction (41).

Bevantolol

Bevantolol is a β_1-selective blocker with partial α-adrenoceptor agonist activity (42) and has been reported not to cause the side effect of cold extremities, but experience is so far insufficient to provide a proper assessment of its clinical efficacy.

Carvedilol

Another new β-blocker with a novel additional action is carvedilol (43,44). It is a nonselective β-blocker with vasodilator and antioxidant properties. One thousand and ninety-four patients with chronic heart failure (ejection fraction ≤ 0.35), who were all on diuretic therapy and receiving Ace inhibitors, were administered placebo or carvedilol in addition. After 6 months of follow-up, mortality in the placebo group was 7.8% but only 3.2% in the carvedilol group. There was no obvious difference in quality of life, raising the question whether such a trial would justify a policy of treating all chronic heart failure patients with beta-blockers in addition to other therapy (45).

In conclusion, the β-blockers have proved over the past thirty years to be safe and efficacious, without serious side effects. Their disadvantages stem from their therapeutic action by taking the edge off peak performance and causing fatigue and reduced sexual performance in some men. Other side effects (vivid dreams, cold extremities) are seldom sufficiently severe to warrant cessation of treatment. The reduction of mortality in post-MI patients may be due to falls in heart rate and blood pressure, but a Class 2 antiarrhythmic action may also be involved.

REFERENCES

1. Pluim BM, Chin JC, DeRoost A, Doornbost J, Siebelink H-MJ, Van der Laarse A, Vliegen HW, Lamerichs RMJN, Bruschke AVG, Van der Waal EE. Cardiac anatomy, function and metabolism in elite cyclists assessed by magnetic resonance imaging and spectroscopy. Eur Heart J 1996; 17:1271–1278.
2. Hutter OF, Trautwein W. Vagal and sympathetic effects on the pacemaker fibers in the sinus venosus of the heart. J Gen Physiol 1956; 39:715–733.
3. Yang Z-K, Boyett MR, Janvier NC, McMorn SO, Shui Z, Karim F. Regional differences in the negative inotropic effect of acetylcholine within the canine ventricle. J Physiol 1996; 492:789–806.
4. Giles W, Noble SJ. Changes in membrane currents in bullfrog atrium produced by acetylcholine. J Physiol 1976; 261:103–123.
5. Brown HF, Giles W, Noble SJ. Cholinergic inhibition of the frog sinus venosus. J Physiol 1977; 267:38–39P.
6. Janvier NC, Boyett MR, Orchard CH, Yang Z-K. The use of the action potential clamp technique to study the negative inotropic effect of acetylcholine on isolated ferret ventricular cells. J Physiol 1995; 487:127–128P.
7. Brock JA, Cunnane TC. Inhibition of purinergic transmission by prostaglandin E_1 and E_2 in the guinea pig vas deferens: an electrophysiological study. Br J Pharmacol 1996; 118:776–782.
8. Todorov LD, Mihaylova-Todorova S, Craviso GL, Bjur RA, Westfall DP. Evidence for the differential release of the co-transmitters ATP and noradrenaline from sympathetic nerves of the guinea-pig vas deferens. J Physiol 1996; 496:731–748.
9. Bryant SM, Hart G. Effects of neuropeptide Y on L-type calcium current in guinea-pig ventricular myocytes. Br J Pharmacol 1996; 118:1455–1460.
10. Sosunov EA, Anyukhovsky EP, Rosen MR. Effects of exogenous neuropeptide Y on automaticity of isolated Purkinje fibers and atrium. J Mol Cell Cardiol 1996; 28:967–975.
11. Saxena PR. Interaction between the renin-angiotensin-aldosterone and sympathetic nervous systems. J Cardiovasc Pharmacol 1992; 19:S80–S88.
12. Richer C, Doussau MP, Giudicelli JF. Perinopril, a new converting enzyme inhibitor: systemic and regional hemodynamics and sympathetic inhibitory effects in spontaneously hypertensive rats. J Cardiovasc Pharmacol 1986; 8:346–357.
13. Steinfath M, Lavicky J, Schmitz W, Scholz H, Döring V, Kalmár P. Regional distribution of β_1- and β_2-adrenoceptors in the failing and non-failing human heart. Eur J Clin Pharmacol 1992; 42:607–612.
14. Kaumann AJ. (-)-CGP12177-induced increase of human atrial contraction through a putative third β-adrenoceptor. Br J Pharmacol 1996; 117:93–98.

15. Kaumann A, Molenaar P. Differences between the third cardiac β-adrenoceptor and the colonic β₃-adrenoceptor in the rat. *Br J Pharmacol* 1996; 118:2085–2098.

16. Dukes ID, Vaughan Williams EM. Effects of selective α₁-, α₂-, β₁- and β₂-adrenoceptor stimulation on potentials and contractions in the rabbit heart. *J Physiol* 1984; 355:523–546.

17. Yabana H, Sasaki Y, Narita H, Nagao T. Subcellular fractions of cyclic-AMP and cyclic-AMP-dependent protein kinase, and the positive inotropic effects of selective β₁- and β₂-adrenoceptor agonists in guinea pig hearts. *J Cardiovasc Pharmacol* 1995; 26:893–898.

18. Gadsby DC. β-adrenoceptor agonists increase membrane K±-conductance in cardiac Purkinje fibers. *Nature* 1983; 306:691–693.

19. D'Avila A, Wellens F, Andreis E, Brugada P. At what time are implantable defibrillator shocks delivered? *Eur Heart J* 1995; 16:1231–1233.

20. Pitzalis MV, Mastropasqua F, Massari F, Totaro P, Scrutinio D, Rizzon P. Sleep suppression of ventricular arrhythmias: a predictor of beta-blocking efficacy. *Eur Heart J* 1996; 17:917–925.

21. Duffy JF, Kronauer RE, Czeisler CA. Phase-shifting human circadian rhythms: influence of sleep-timing, social contact and light exposure. *J Physiol* 1996; 495: 289–297.

22. Vaughan Williams EM, Dennis PD, Garnham C. Circadian rhythm of heart rate in the rabbit: prolongation of action potential duration by sustained beta-blockade is not due to associated bradycardia. *Cardiovasc Res* 1986; 20:528–535.

23. Kaila T, Tarssanen L, Scheinin M, Kantola I. The effect of β-blockade on plasma potassium and magnesium: homeostasis during exercise. *Am J Ther* 1996; 3: 435–440.

24. Wesslén L, Påhlson C, Lindquist O, Hjelm E, Gnarpe J, Larsson E, Baandrup U, Eriksson L, Fohlman J, Engstrand L, Linglof T, Nyström-Rosander C, Gnarpe H, Magnius L, Rolf C, Friman G. An increase in sudden unexpected cardiac deaths among young Swedish orienteers during 1979–1992. *Eur Heart J* 1996; 17:902–910.

25. Papp JGy, Vaughan Williams EM. The effect of bretylium on intracellular action potentials in relation to its antiarrhythmic and local anaesthetic activity. *Br J Pharmacol* 1969; 37:380–390.

26. Vaughan Williams EM, Bagwell EE, Singh BN. Cardiospecificity of β-receptor blockade. A comparison of the relative potencies on cardiac and peripheral β-adrenoceptors of propranolol, of practolol and its ortho-substituted isomer, and of oxprenolol and its para-substituted isomer. *Cardiovasc Res* 1973; 7:226–240.

27. The Australian and Swedish pindolol study group. The effect of pindolol on the two year mortality after complicated myocardial infarction. *Eur Heart J* 1983; 4:367–375.

28. Kübler W, Haass M. Cardioprotection: definition, classification, and fundamental principles. *Heart* 1996;75: 330–333.

29. Vaughan Williams EM, Raine AEG, Cabrera AA, Whyte JM. The effects of prolonged β-adrenoceptor blockade on heart weight and cardiac intracellular potentials in rabbits. *Cardiovasc Res* 1975; 9:579–592.

30. Raine AEG, Vaughan Williams EM. Adaptation to prolonged β-blockade of rabbit atrial, Purkinje and ventricular potentials, and of papillary muscle contraction. Time-course of development of, and recovery from, adaptation. *Circ Res* 1981; 48:804–812.

31. Vaughan Williams EM, Hassan MO, Floras JS, Sleight P, Jones JV. Adaptation of hypertensives to treatment with cardio-selective and non-selective beta blockers. Absence of correlation between bradycardia and blood-pressure control, and reduction in the slope of the Q-T/R-R relation. *Br Heart J* 1980; 44:473–487.

32. Lo SSS, Mathias CJ, Sutton MSt.J. QT interval and dispersion in primary autonomic failure. *Heart* 1996; 75:498–501.

33. Birkhead JS, Vaughan Williams EM, Gwilt DJ, Tanqueray A, Cazes C. Heart rate and QT interval in subjects adapted to beta-blockade: bradycardia and hypertension as uncorrelated adaptations. *Cardiovasc Res* 1983; 17:649–655.

34. Inoue H, Zipes DP. Time course of denervation of efferent sympathetic and vagal nerves after occlusion of the coronary artery in the canine heart. *Circ Res* 1988; 62:1111–1120.

35. Raine AEG, Vaughan Williams EM. Electrophysiological basis for the contrasting prophylactic efficacy of acute and prolonged beta-blockade. *Br Heart J* 1978; 40: (Suppl) 71–77.

36. Vaughan Williams EM, Tasgal J, Raine AEG. Morphological changes in the myocardium induced by prolonged β-adrenoceptor blockade. *Br Heart J* 1978; 40:454.

37. Tasgal J, Vaughan Williams EM, The effect of prolonged propranolol administration on myocardial transmural capillary density in young rabbits. *J Phsyiol* 1981; 315:353–367.

38. Singh BN, Vaughan Williams EM. A third class of antiarrhythmic action. Effects on atrial and ventricular intracellular potentials, and other pharmacological actions on cardiac muscle of MJ1999 and AH3474. *Br J Pharmacol* 1970; 39:675–687.

39. Singh BN, Vaughan Williams EM. Local anaesthetic and antiarrhythmic actions of alprenolol relative to its effect on intracellular potentials and other properties of isolated cardiac muscle. *Br J Pharmacol* 1970; 38:749–757.

40. Singh BN, Vaughan Williams EM. The effect on cardiac muscle of the β-adrenoceptor blocking drugs INPEA and LB46 in relation to their local anaesthetic action on nerve. *Br J Pharmacol* 1971; 43:10–22.

41. Vaughan Williams EM, Millar JS, Campbell TJ. Electrophysiological effects of labetolol on rabbit atrial, ventricular and Purkinje cells, in normoxia and hypoxia. *Cardiovasc Res* 1982; 16:233–239.

42. Vaughan Williams EM. Bevantolol: a beta₁-adrenoceptor antagonist with unique additional properties. *J Clin Pharmacol* 1987; 27:450–460.

43. Raftery EB. Vasodilating beta-blockers in heart failure. *Eur Heart J* 1995; 16:(Suppl T) 32–37.

44. Feuerstein GZ, Ruffolo RR. Carvedilol, a novel vasodilating beta-blocker with the potential for cardiovascular organ protection. *Eur Heart J* 1996; 17:(Suppl B) 24–29.

45. Cohn JN. Slowing the progress of heart failure. *Eur Heart J* 1996; 17:1609–1611.

4

Class 3 Antiarrhythmic Action

E. M. Vaughan Williams

During the last couple of centuries *Homo sapiens* has begun to unveil the complexity and subtlety of the machinery that has evolved over 3.5 billion years, enabling multifarious species to survive in their specific habitats. Many basic biochemical pathways are shared by most organisms, but higher functions still elude analysis. A tennis player returning a fast service employs visual and muscular mechanisms far more sophisticated than any defense system tracking an incoming missile. Investigators face a dual challenge; they may discover a molecule the function of which is unclear (eg, neuropeptide Y), or they can deduce that a control factor must exist (eg, for adjusting circadian sympathetic activity to light) without knowing what it is (melatonin?). The spread of excitation in the ventricle is finely co-ordinated, travelling down the Purkinje system at 3 m.s^{-1}, then progressing from three widely separated sites initiated simultaneously, and spreading outwards in concentric leaflets at 0.5 m.s^{-1} from endocardium to epicardium (1).

Since all the ventricular cells are electrically interconnected through gap junctions, if the normal orderly progression of excitation is interrupted or confused by an appropriate pattern of electrical stimuli, ventricular tachycardia (VT) or fibrillation (VF) can be precipitated in perfectly normal hearts. The action potential duration (APD) of epicardial cells is shorter than that of endocardial cells, and this in turn is shorter than that of the terminal Purkinje cells. The cells that depolarize last repolarize first, proximal cells always remaining refractory until their distal neighbors have repolarized, making reflection or reentry of excitation impossible (Fig. 1). Whether the action potential duration (APD) is genetically determined to be permanently differentiated is unknown. Would an epicardial cell transposed to an endocardial position retain its short APD or adapt to its situation and keep in phase with its new surroundings?

Disparity of APD is arrhythmogenic. QT-interval dispersion predicts cardiac death in patients with vascular disease (2). Monophasic action potential (MAP) recordings from different sites in the heart have revealed that disparity of APD is associated with arrhythmia. A link between the dispersion and the inducibility of a monophasic VT was found, which suggests that the increased dispersion plays an important role in the genesis of a *monomorphic* VT (3). In another recent study with MAP recording it was concluded that increased dispersion of ventricular repolarization is one of the underlying mechanisms accounting for myocardial vulnerability to ventricular arrhythmias. Repolarization disturbance is important for the genesis of *polymorphic* tachycardia/fibrillation (4). MAP recording is invasive and not widely available, and much effort has gone into devising computer programs for estimating QT-interval dispersion noninvasively from the ECG. Several different measurement techniques have been assessed by McLaughlin et al. (5).

The equilibrium potential for sodium and calcium changes rapidly between systole and diastole, but is always positive. Although the *average* [Na]$_i$ may be about 10 mM, at the peak of the action potential it rises rapidly, and diffusion into the cell interior is believed

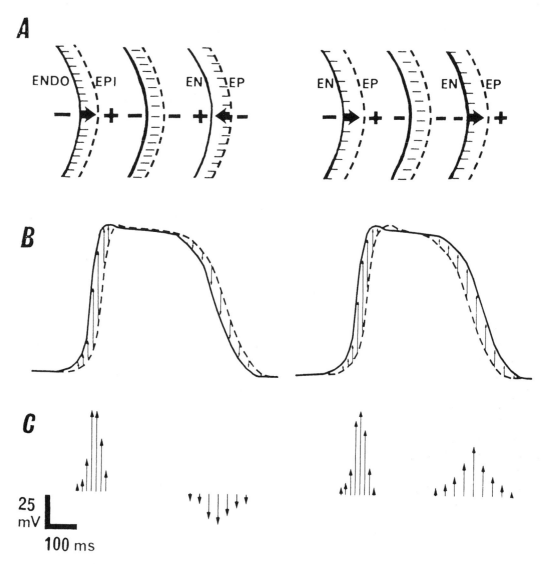

FIG. 1. *Left.* **A.** Diagram of LV (Left Ventricular) free wall. Endocardium (*solid line*) depolarizes first, and current flows from resting epicardium (*dotted line*). When this also depolarizes, the whole wall is equipotential, and no current flows. If APD was the same throughout, endocardium would repolarize while epicardium was still depolarized, making it possible for the latter to reexcite the endocardium (reflection). **B.** Diagram of endocardial and epicardial action potentials, on the assumption that APD was equal. Arrows depict endo/epi potential differences. **C.** Replot of arrows on horizontal baseline. Endo/epi potential differences during depolarization (R-wave) and repolarization (T-wave) are opposite in sign (discordant). *Right.* **A, B,** and **C** as above, but with assumption that epi repolarizes before endo, making reflection impossible. R and T waves are now concordant.

to be restricted by a subsarcolemmal barrier. The $[Ca]_i$ has to fall during diastole to $<10^{-7}$ M to permit muscle relaxation, but rises to $>10^{-6}$ M at the peak of contraction. The chloride equilibrium potential is less volatile, and is about -30 mV. Thus, apart from active repolarization by the Na/K pump, for the cell to repolarize passively to a level more negative than -30 mV, the permeability to potassium must greatly exceed that of the other ions.

POTASSIUM CHANNELS

Delayed Rectification

The introduction of practolol as a β-blocker without local anesthetic activity led by accident to the discovery of the distinction between β_1-adrenoceptors, which were blocked by the drug, and β_2-adrenoceptors, which were not. Likewise, the repolarizing current, I_K, was found to have a rapidly activating component, I_{Kr}, which was blocked by sotalol and by another methane sulfonamide derivative, E4031, and a more slowly activating component, I_{Ks}, which was less sensitive to block (6). A specific protein minK or IsK "underlies I_{Ks} in guinea pig heart. A gene (human-ether-a-gogo related gene, or HERG) coding for channels resembling I_{Kr} has been identified in the human heart" (7). In apparently normal human myocytes, obtained at surgery from the right free ventricular wall, clamp studies confirmed that, as in guinea pig and canine tissue, "the E4031-sensitive component (I_{Kr}) activates more rapidly, at a more negative potential, and with a more steep slope factor than does the E4031-resistant component (I_{Ks}) and shows inward rectification. Like I_{Ks} in guinea pig ventricle, E4031-resistant current in human ventricle is effectively inhibited by the diuretic agent indapamide." Another diuretic, triamterine, was also found to block I_K in guinea pig ventricular myocytes, with selectivity for I_{Ks} over I_{Kr} (8). In human myocytes obtained from patients with ischemic or dilated cardiomyopathy, however, action potentials were exceptionally prolonged (>1.2 seconds), and single-channel studies revealed that the I_{Ks} component of I_K appeared to be absent, providing direct evidence that diseased myocardium undergoes a selective prolongation of APD, creating an arrhythmogenic disparity of repolarization (9).

Ultra-rapid Delayed Rectification

A third type of delayed rectification in canine atrial cells has been described, though the term *delayed* is hardly appropriate because at 35°C the activation time constant was 2.5 ms in response to depolarization to -10 mV, and 0.33ms after a step to +60 mV. The current was activated between -70mV and -50 mV, and exhibited characteristics of a potassium current (from reversal potential and effect of changes in $[K^+]_o$). Inactivation was much slower and biexponential, with time constants of 609 ± 90 and 5565 ± 676 ms. Block of the current, by 4-AP and TEA, prolonged APD. The current was similar to that carried by a cloned KV 3.1 channel. (10)

Inward Rectification

The resting potential in Purkinje and ventricular cells is >-90mV, close to E_K, and is stabilized by high conductance through the inward rectifier channels. If extracellular potassium concentration $[K]_o$ is reduced, E_K becomes more negative and the resting potential is positive to E_K, causing K conductance to fall (ie, the channel rectifies) (Chapters 1, 7). The resting potential is consequently less stable, accounting for the greater probability of arrhythmias in hypokalemia. When the cell depolarizes, inward rectification assists maintenance of the plateau, at the end of which repolarizing current flows through I_K channels. As the potential approaches E_K, the inward rectifier current, I_{K1}, cuts in again to reinforce the stability of the resting potential. Na^+ channels recover from inactivation at potentials negative to -50 mV, and the effective refractory period (ERP) is determined at the point (about -65mV) at which sufficient Na^+ current is available to sustain a propagated impulse. I_{K1} is not fully reactivated, however, and for a brief vulnerable period thereafter an applied stimulus can most easily initiate an arrhythmia if there is a susceptible substrate. At electrophysiological study (EPS) stimuli programmed to fall just outside the ERP of one or more successive action potentials can most effectively test for such susceptibility.

In pacemaking cells of the sinoatrial node (SAN) the resting potential is unlocked, as it were, from E_K by the absence of I_{K1} channels.

A drug that selectively blocked I_{K1} would not be expected to affect the plateau, but to prolong the tail of the action potential. Whether such a drug would be antiarrhythmic is questionable, however, because, although it would exert a Class 3 effect, the resting potential would be less stable and more vulnerable to invasion by an ectopic impulse. RP58866 was reported to block I_{K1} selectively. In rats it increased the QT-interval, without reverse frequency dependence, and reduced the incidence of ischemia-induced VF from 100% to 50%, 17%, and 0% at concentrations of 1, 3, and 10 μM, respectively. In rabbits and primates it was less effective but caused some reduction in ischemia-reperfusion-induced arrhythmias. Strict selectivity for I_{K1}, however, seems doubtful, because RP58866 caused bradycardia in the SAN, in which I_{K1} channels are absent (11).

Transient Outward Current

In epicardial cells and some Purkinje cells a brief repolarization or "notch" is observed between the fast phase of depolarization and the plateau. Carmeliet (12) concluded that chloride ions contribute little to the total membrane conductance at the resting potential, but become important as electrical charge carriers when the membrane is depolarized. Dudel et al. described an outward current, which was reduced by removal of extracellular chloride, and which they called dynamic current (13). Fozzard and Hiraoka (14) also concluded that a positive dynamic current governed by Hodgkin-Huxley variables was carried by Cl⁻ ions, and McAllister et al. (15) accepted it as a chloride current in a reconstruction of the cardiac action potential, calling it I_{qr}.

Kenyon and Gibbons reinvestigated what they now called early outward current in Purkinje cells and concluded that it was carried by potassium ions, because it was reduced by substances blocking K⁺ channels (eg, 4-AP), and it was not reduced in solutions containing only 8.6% of normal Cl⁻ concentration (16,17). Carmeliet and Verdonck, however, observed

that substitution of Cl⁻ by other anions reduced K⁺ permeability by 38%, which might explain why the transient current was reduced in low Cl⁻ even if carried by K⁺ ions (18).

Commenting on evidence that transient outward current (now designated I_{TO}) was activated by calcium, Ten Eick et al. observed "I_{TO} appears to have both Ca^{2+}-independent and Ca^{2+}-dependent (I_{TO-Ca}) components. Until recently it was thought that I_{TO-Ca} was carried by K⁺. However, I_{TO-Ca} has been shown to exhibit a Cl⁻-dependence, suggesting that Cl⁻ is the principal charge carrier for the Ca^{2+} dependent component" (19).

None of these experiments explains the function of I_{TO}. It is present mainly in atrial, proximal Purkinje, and epicardial cells, all of which have shorter APDs than the ventricle, in which I_{TO} is small and rapidly inactivated (20). Since it is a transient current, it cannot be directly responsible for the shortening of APD. It seems clear that I_{TO} is carried by both K⁺ and Cl⁻ ions, the Ca-independent component being blocked by 4-AP and the Ca-dependent component by the anion channel blocker DIDS (21). In the human atrium I_{TO} differs in several respects from I_{TO} in the epicardium. In the atrium I_{TO} is inactivated much more slowly and has been called I_{SO} (22). To predict what overall effect on the heart would be exerted by a selective blocker of one or more components of I_{TO} is obviously premature.

SODIUM-ACTIVATED POTASSIUM CURRENT

An outward K⁺ current (K_{Na}) is activated "if the intracellular sodium concentration $[Na]_i$ is raised to 30mM–100 mM in membrane patches of ventricular myocytes" (23). Since $[Na]_i$ would never reach such levels in a living heart, such experiments could be dismissed as an artefact of no significance. If, however, a subsarcolemmal barrier exists to diffusion of Na⁺ ions into the cell interior, a *local* concentration of sodium could be created, by rapid stimulation, in the neighborhood of such K_{Na} channels high enough to activate them, especially if the removal of Na^+ were restricted by inhibition of the Na/K pump. Activation of K_{Na} could account for the shortening of APD by cardiac glycosides.

ATP-REGULATED POTASSIUM CURRENT (I_{K-ATP})

Hypoxia causes an immediate depression of cardiac contractions, accompanied by a shortening of APD. Repeated periods of mild hypoxia (20% O_2) in the presence of glucose, which would maintain a high level (>6 mM) of intracellular ATP (24), were alternated with periods of normoxia, during which contraction and APD completely recovered. In each succeeding period of hypoxia the onset of depression of contractions and shortened APD was more rapid and profound and was associated with a progressive depletion of glycogen. The effect was exacerbated by free fatty acids (FFA), and it was concluded that FFA accelerated glycolysis (25).

Hypoxia was shown many years ago to increase potassium efflux (26). If ATP synthesis is inhibited by cyanide in guinea pig ventricular cells, no effect on K efflux is observed until [ATP]$_i$ falls below 2 mM, and a substantial effect required ATP < 1mM. "It is concluded that ATP-regulated K^+ channels are responsible for the increase in outward current and the shortening of action potential duration in various anoxic conditions" (27). To open ATP-regulated channels in ischemia, in contrast to cyanide, requires total ischemia for many minutes in the absence of carbohydrate substrate. [ATP]$_i$ must fall to micromolar levels and the channels close immediately on the readmission of oxygen.

ATP-regulated channels primarily control functions at extracardiac sites, in the pancreas, and in smooth muscle of the vasculature. I_{K-ATP} hyperpolarizes the cells, causing inhibition of insulin secretion and relaxation of vessels respectively. Sulfonyl-ureas block the channels, and insulin is released. Glyburide blocks ATP-regulated channels in cardiac muscle also, but has no effect on normal function. Blockade by gliburide or glibenclamide of a current does not prove that the current was I_{K-ATP}; other currents could also be blocked.

If opening of ATP-regulated channels is responsible for APD shortening in hypoxia, it is necessary to explain why the channels remain open in spite of the presence of oxygen and glucose and an [ATP]$_i$ > 6 mM. It has been proposed that hypoxia activates an "endogenous" K-channel opener (28). Alternatively, "the sensitivity of the channels towards internal ATP is drastically reduced during metabolic stress" (29). It was demonstrated that products of ischemia did reduce the ATP-sensitivity of ATP-regulated channels to ATP, but the K_m was increased from 49 μM to only 226 μM (30), so that such a small change could hardly be significant in the presence of 6 mM ATP. Another possibility is that a subsarcolemmal barrier exists restricting access of cystosolically produced ATP to the ATP-regulated channels. As noted above, the effect of hypoxia on APD increased as glycogen was depleted, and it was observed that "the ATP which regulates channel activity is produced by membrane *delimited glycolysis*" (29).

Thus at present it seems that mild hypoxia in the presence of glucose may shorten APD by a mechanism other than, or additional to, the activation of I_{K-ATP}, a possibility supported by the fact that sulfonyl-ureas only partially and transiently counteract the effects of hypoxia or ischemia (29).

POTASSIUM CURRENT ACTIVATED BY ACETYLOCHOLINE (I_{K-ACH})

This current, already described, has sometimes been regarded as similar to, or interchangeable with, I_{K-ATP}, as if Ach were just another K-channel-opener. It has been shown, however, that glyburide does not alter I_{K-Ach}. Adenosine, on the other hand, may activate both I_{K-Ach} and I_{K-ATP}, its effect being reduced but not abolished by glyburide (31). There is evidence that adenosine, like Ach, can inhibit noradrenaline release from postganglionic sympathetic neurones (32).

MOLECULAR STRUCTURE OF POTASSIUM CHANNELS

A mutation in *Drosophila melanogaster* causing it to shake under ether anesthesia was traced to an alteration of a specific K^+ channel (shaker). Analysis of cDNA sequences revealed homologies to the shaker gene in rat and human hearts. Six putative membrane-spanning columns line the pore region through which K^+ ions traverse the membrane, and this central core is highly conserved among numerous isoforms (33). Since the cloning of the first inwardly rectifying channel in 1993, a family of related clones has been isolated. Classic inward rectification is caused by a very steeply voltage-dependent block of the channels by polyamines, with an additional contribution by Mg^{2+} ions, and relief of this plugging at negative voltages is the activation process. Alterations in the cellular concentration of polyamines could provide a mechanism for controlling the degree of rectification of K^+ channels (34). Six different K^+ channels have been cloned from rat and human heart, and all have been functionally characterized in either *Xenopus* oocytes or mammalian tissue culture systems (35).

When the molecular basis for ion selectivity, voltage-gating, and receptor affinity has been determined, it may be possible to synthesize drugs to fit specific parts of individual channels. Ultimately, however, any drug, whatever the origin of its design, will have to be submitted to clinical trial.

EFFECTS OF INDIVIDUAL CLASS 3 AGENTS

Amiodarone

Amyl nitrite was originally thought to exert its beneficial effect in angina pectoris by dilating coronary arteries, and a pharmacological objective was to produce more potent and selective coronary dilators. Narrow coronaries with thick walls cannot dilate sufficiently and must be replaced (coronary bypass) or have their lumen enlarged (angioplasty). Vasodilators reduce oxygen demand by relief of pressure afterload or preload or both. Amiodarone, a benzfuran vasodilator, was but one of a large series of antianginal compounds synthesized and studied at Labaz in Brussels by a brilliant team that included Guy Deltour, Robert Charlier, and their consultant ZM Bacq.

CLASS 2 ACTION

The beta-blockers were introduced in the early sixties (pronethalol, 1962; propranolol, 1964) and were antianginal. The Labaz group considered, however, that it would be preferable to induce only a partial antagonism to adrenergic stimulation, for fear of compromising a possibly vital positive inotropic sympathetic support, and amiodarone reduces by about one-third sympathetic effects mediated by both α- and β-receptors. It is not surprising, therefore, that acute administration of amiodarone, if injected too rapidly, may cause bradycardia, hypotension, and prolonged AV (Atrio Ventricular) conduction, because it is a vasodilator and antagonizes the high sympathetic tone which is often prevalent in patients with cardiac disease.

Charlier found that amiodarone was antiarrhythmic in several standard laboratory models of arrhythmias, such as those induced acutely by aconitine, calcium chloride, chloroform, coronary ligation, and so forth. It was not, however, particularly potent, having about one-sixth the activity of quinidine, and its mode of action was unclear. Charlier had demonstrated that the antisympathetic action of amiodarone was *not* due to adrenoceptor blockade (36), but some authors have persisted in referring to the beta-adrenergic blocking properties of amiodarone, for example, to explain why epinephrine reversed effects of quinidine but not of amiodarone (37). The antisympathetic action may be partly post-synaptic and intracellular, but Bacq et al. proved that amiodarone had in addition a presynaptic action, inhibiting the release of noradrenaline from postganglionic sympathetic neurones (38). This explains why amiodarone antagonizes effects mediated by both α- and β-adrenoceptors. There is no inhibition of noradrenaline uptake. Amiodarone has a negligible negative inotropic action, but vasodilatation and a reflex response may increase cardiac output (39). *In vitro* amiodarone causes bradycardia, but in conscious animals heart rate may be unchanged or increase in response to acute hypotension.

CLASS 3 ACTION

Treatment of rabbits for 3-6 weeks with daily intra-peritoneal injections of amiodarone (in distilled H_2O) caused a prolongation of APD in atria and ventricles. Amiodarone contains iodine, which is split off as iodide, and since it had recently been shown that thyroidectomy prolonged APD (see Chapter 1, Ref. 16), it was suspected that an antithyroid effect might be involved. Administration of iodide had no effect on APD, however, and amiodarone did not antagonize the action of thyroxine (see Chapter 1, Ref. 17). The effect on APD took several days to develop, reaching a maximum at 3 weeks. In acute experiments on isolated guinea-pig ventricular cells, with both whole-cell and single-channel patch-clamp technique, amiodarone reduced inward rectifier current, I_{K1}, but high concentrations were required (10 μM to 20 μM) to block the channels (40). In contrast, Varró et al. reported that neither acute nor chronic administration of amiodarone altered I_{K1} in isolated rabbit cardiac myocytes, but acute and chronic treatment reduced the delayed rectifier current, I_K. There was no effect on I_{TO} acutely, but some depression after chronic treatment (41). Subdividing I_K, amiodarone blocked the slowly activating component (resistant to La^{3+}) in guinea pig ventricular myocytes, but not the rapidly activating (La^{3+}-sensitive) component. "Amiodarone block did not increase on a step-decrease in membrane potential. These findings may in part explain the low propensity for amiodarone to cause repolarization-related arrhythmias" (42).

In human right ventricle the effect of amiodarone on APD does not exhibit reverse frequency dependence. "Prolongation of APD

after amiodarone was independent of pacing rate; eg, 12% ± 9% at cycle length (CL) 700ms, and 11% ± 6% at CL 350ms (43). Amiodarone induced a uniform lengthening of APD after abrupt changes of cycle length.

CLASS 1 ACTION

Long-term treatment of rabbits with amiodarone induced a minor reduction of V_{max} in sinus rhythm (see Chapter 1, Ref. 17). Mason et al. reported (44) that amiodarone blocked inactivated sodium channels, but the pharmacological profile of amiodarone is not that of a Class 1 drug, and its clinical antiarrhythmic efficacy is associated with prolongation of QT interval. In patients treated for 2 to 10 weeks there was no change in QRS duration on pacing at a cycle length (CL of 700 ms, which is consistent with the lack of Class 1 action in rabbits in sinus rhythm. When the patients' hearts were paced at higher frequencies, however, QRS duration increased, becoming significantly longer at CL of 300 ms. The onset time-constant was very short, 306 ± 122 ms, so that the Class 1 action of amiodarone is in the 1b category (45). The effect of amiodarone on sodium current was confirmed by voltage-clamp experiments on ventricular myocytes taken from 2- to 5-day-old rabbits (46).

CLASS 4 ACTION

In vitro, amiodarone slows the frequency of the SAN and delays AV conduction. In humans the long-term effect of amiodarone is bradycardia, but acutely there may be a reflex response to hypotension on intravenous (IV) administration. AV conduction is delayed. Amiodarone has a negligible negative inotropic action, and the effect on the SAN could be accounted for by the prolongation of APD, and on both nodes by the Class 2 action. In voltage-clamp studies on isolated cardiac myocytes amiodarone reduced I_{Ca} (41), so that it is possible that a Class 4 action contributes to the effects on the nodes. Verapamil sharpens the cardiac action potential foot (representing passive flow of depolarizing current through gap junctions from the approaching active region), and it was suggested that longitudinal resistance could be reduced by lower $[Ca^{2+}]_i$ (47). It has now been proposed, on the basis of calculations of transverse and longitudinal resistivity from measurements of V_{max} and conduction velocity in strips of canine myocardium, that amiodarone may improve conduction by increasing gap junction conductance (48,49). The effect of amiodarone on calcium current cannot be large, because it would be reflected in a proportionate negative inotropic action, especially in combination with the Class 2 action. It is possible, however, that amiodarone selectively inhibits I_{Ca} in the SAN and AVN, and perhaps at gap junctions.

THYROID INTERACTION

In the presence of a preexisting disturbance of thyroid function amiodarone may exacerbate thyrotoxicosis or hypothyroidism, and patients with a personal or family history of abnormality should be excluded from treatment or be closely monitored. Amiodarone inhibits the enzyme 5'deiodinase, and long-term treatment raises serum levels of total T_4 and free thyroxine and of reverse T_3. Total T_3 and free T_3 is either normal or decreased, and TSH remains unchanged (50). Unless, therefore, there is an underlying abnormality, amiodarone does not seriously disturb thyroid function. Nevertheless, the similarity between several of the effects of amiodarone and of hypothyroidism (bradycardia, reduced O_2 uptake, prolonged APD, diminished response to sympathetic stimulation) and the slow onset of the full therapeutic effect has led to an unproven suspicion that a selective interference within cardiac cells somewhere along a common pathway may be involved (51).

CLINICAL EFFICACY

The antiarrhythmic action of amiodarone was originally demonstrated in numerous animal models and was later confirmed in patients with supraventricular and ventricular arrhythmias. Amiodarone had been in use as an antianginal remedy for years already, and long-term oral treatment was customary because of the slow onset of the full effect. The antiarrhythmic action, however, was observed acutely after intravenous administration in animals (52) and in patients (53). Since CAST, interest has intensified in Class 3 antiarrhythmic action, and several long-term trials of amiodarone treatment in post-MI (Myocardial Infarction) and other patients have been completed, with favorable results, or are ongoing (54). In the Basel study of post-MI patients "the probability of survival of patients given amiodarone was significantly greater than that of control patients" (55). The benefit was less apparent in subjects with depressed left ventricular function (56). In a Canadian study of post-MI patients followed for up to 2 years (mean, 20 months) arrhythmic death occurred in 6% of amiodarone-treated patients and in 14% of those on placebo. Ventricular premature depolarizations (VPD) were reduced. "Amiodarone, in moderate loading and maintenance dosages, with *adjustments* in response to plasma levels, VPD suppression and side effects, results in effective VPD suppression and acceptable levels of toxicity" (57). The adjustment of dosage to suit individual cases may be important in long-term trials. A double-blind, placebo-controlled Polish trial of amiodarone in post-MI patients "demonstrated a significant reduction in cardiac mortality and ventricular arrhythmias" (58). A five-year follow-up of 589 patients treated with amiodarone confirmed that "despite its side effect profile, amiodarone is an effective and reasonably well tolerated antiarrhythmic drug" (59).

Even after cessation of treatment at the end of a year, "the beneficial effect of amiodarone persists for several years. The rate of sudden death and all cardiac death is low (1.6% and 4.1%, respectively) during the late follow-up and therefore may not warrant further therapy" (60).

In addition to long-term prophylactic treatment, amiodarone can be used acutely in a variety of arrhythmias, supraventricular and sustained ventricular tachycardia or fibrillation (61). Short-term control of serious arrhythmias can be exerted by IV administration, though the full therapeutic effect may not be observed for days (62). Even oral amiodarone has an acute effect, if a sufficient initial loading dose is administered. "Oral amiodarone given in a loading dose produces rapid and dramatic reductions in spontaneous ventricular arrhythmias within 72 hours" (63).

AMIODARONE IN ISCHEMIA

Ischemia shortens APD by increasing potassium conductance. In acute experiments on isolated tissues, Cobbe et al. found that the effect of amiodarone in prolonging APD was abolished by *total* anoxia, with the implication that the Class 3 action would be lost in arrhythmias associated with ischemic heart disease (64). Even if APD did not lengthen in the ischemic zone, it would still do so in the surrounding normoxic region, which would be protected from invasion. In ischemia induced by block of a coronary artery, however, anoxia is never total, because of collateral blood flow, and in a model of ischemia produced by coronary ligation in anesthetized dogs, the exact opposite of Cobbe's results was demonstrated. Amiodarone lengthened APD *more* in the ischemic than in the normally perfused zone, so that the disparity of APD between the two regions was *reduced* (65).

SIDE EFFECTS OF AMIODARONE

Amiodarone is firmly bound to many tissues and accumulates over a prolonged period (months). The serum concentration of amiodarone may not be a reliable guide to optimum dosage, and it has been suggested that "the peak antiarrhythmic activity of the drug correlates better with the time course of desethyl-amiodarone concentration than with amiodarone level.... The extensive tissue deposition may be a more important reason for toxicity than serum concentration of unchanged drug or metabolite" (66).

In some patients treated with amiodarone, microdeposits have been observed in the cornea, which at one time caused alarm; but in a survey of 8000 subjects they occurred in only 2%. There was no impairment of vision, and the deposits disappeared on cessation of treatment (50).

Amiodarone may induce cutaneous photosensitivity, especially in fair-skinned subjects, leading eventually to a slate-blue discoloration of unprotected skin. The pigmentation did not appear until the second year of treatment. The wave-length of light involved is longer than the ultraviolet (UV) causing sunburn.

Pulmonary fibrosis has been reported, in less than 1% of long term treatments, with an estimated incidence of 0.1 to 0.4 per 1000 patient years. The pulmonary involvement is an idiosyncratic reaction, with no relationship between mean cumulative dose per kg body weight and pulmonary toxicity (50).

Peripheral neuropathy associated with amiodarone has not been proven to be drug related. In three retrospective studies involving 2000 patients in Europe, there were no cases of peripheral neuropathy; but in nerve biopsies abnormalities were detected by light and electron microscopy.

Risk factors for the incidence of noncardiovascular adverse effects of amiodarone in 44 consecutive patients were analyzed by Tisdale et al. (67). Dermatological side effects were more common in young than in old subjects. No specific risk factors for thyroid effects or pulmonary fibrosis were detected.

The incidence of amiodarone-induced exacerbation of arrhythmias (proarrhythmia) is very low in comparison with other antiarrhythmic agents, apart from β-blockers.

Sotalol

In 1970 the β-blocker MJ1999, an ethanolamine with methyl-sulfonamide side chain (Fig. 2), was observed to prolong APD in rabbit atria and ventricles (see Chapter 3, Ref. 35). This property attracted little attention for a decade, but the success of amiodarone as an antiarrhythmic agent drew attention to the

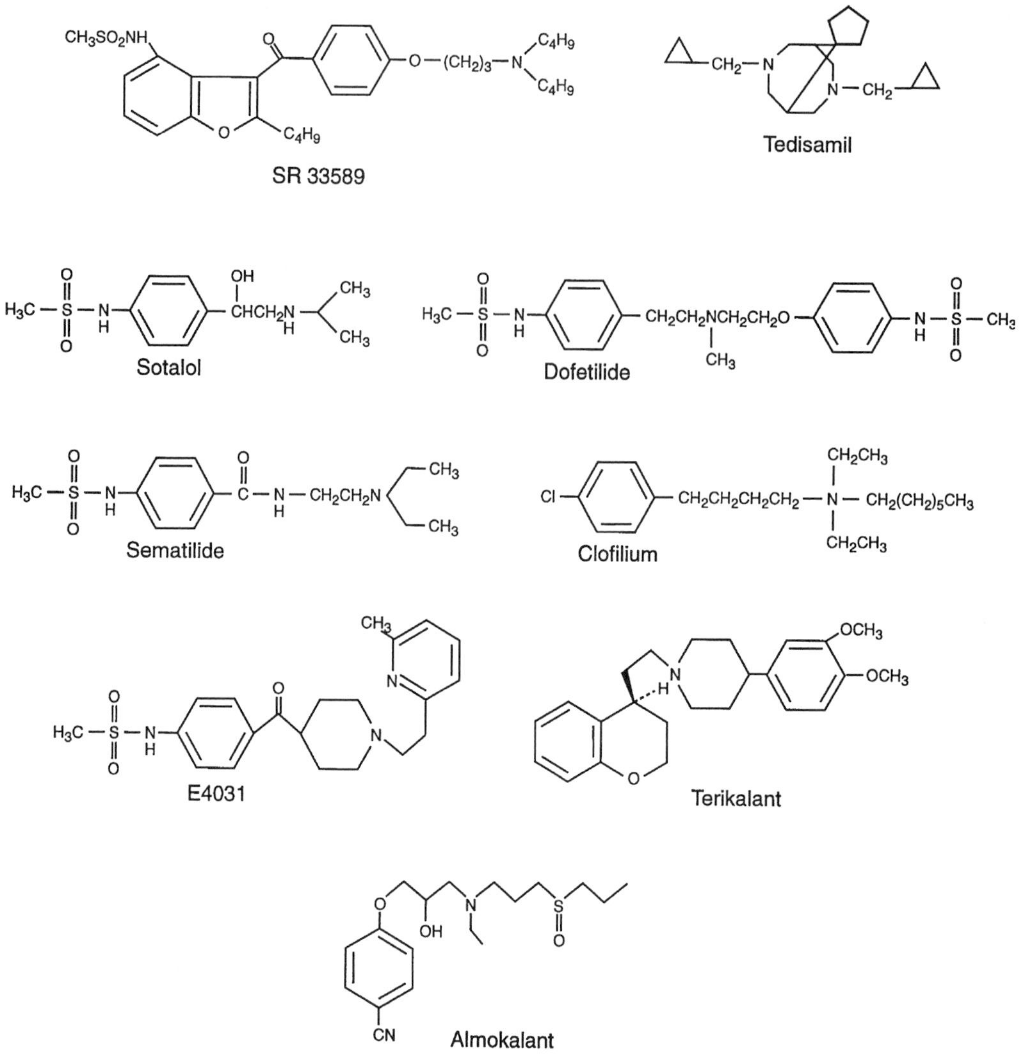

FIG. 2. Structures of compounds described in the text.

possibility that the Class 3 action of MJ1999, now called sotalol, might yield a clinical benefit. Nathan et al. provided evidence that sotalol had an antiarrhythmic action *additional* to what was attributable to its β-blocking action (68), and Bennett demonstrated an acute prolongation of myocardial refractoriness (69). The Class 3 action of sotalol was subsequently confirmed in various cardiac preparations and species (70). In voltage-clamp experiments on myocytes of guinea pigs and rabbits, sotalol 60 μM depressed the delayed rectifier current, I_K, without effect on I_{K1}, I_{TO} or on sodium or calcium currents (71), but Escande and Henry quoted evidence that sotalol blocked all three K currents (33). Species differences have to be taken into account, but there is consensus that sotalol blocks I_K .

Reverse Frequency Dependence

Monophasic action potential (MAP) recording from human right ventricle after 2 mg.kg^{-1} d-sotalol IV caused APD prolongation of 11% at a pacing cycle length of 700ms, but of 5% at CL of 350ms (d- and l-sotalol are equipotent for effect on APD) (72). In anesthetized dogs, 8 to 12 days after ligation of the left anterior descending coronary artery, MAP$_{90}$ in the normal ventricle was increased by 1.5 mg.kg^{-1} dl-sotalol IV from 184 to 225 ms, and in the infarcted zone from 185 to 226 ms. An extra stimulus applied during pacing at short coupling intervals reduced the APD lengthening effect of sotalol in the normal myocardium, but reduced it less in the infarcted zone. This result is again the opposite of that reported by Cobbe et al. in totally anoxic isolated tissue and emphasizes the difference between long-surviving cells in an infarcted region (in which no shortening of APD was observed), and cells made acutely ischemic. The reverse frequency dependence of the effect of sotalol on APD was demonstrated in atria of anesthetized dogs by Wang et al. (73).

Clinical Efficacy of Sotalol

Abundant evidence has accumulated since CAST of the antiarrhythmic efficacy of sotalol. In 114 patients with chronic ventricular premature contractions (VPC) it was concluded that "sotalol is an efficacious antiarrhythmic drug for VPC suppression" (74). Proarrhythmia was observed in three patients on sotalol and in two on placebo. Sustained ventricular tachycardia (SVT) also responded to oral sotalol. "Sotalol is effective (45%) against SVT induction at moderate doses, and is well tolerated over a long term in the setting of remote MI. Its electrophysiologic effects measured at invasive testing are not predictive of efficacy against VT induction" (75). In another study oral sotalol controlled 60% of cases of symptomatic VT in patients with coronary artery disease (76). Rapid IV infusion of sotalol in 109 patients with sustained VT was reported to be safe, even in subjects with a low ejection fraction, and was effective within minutes. Hypotension was observed in only two patients, and corrected by volume infusion.

There was no bradycardia, proarrhythmia, or worsening of ventricular function (77).

Sotalol has been successfully used long term for VT refractory to Class 1 drugs. The antiarrhythmic effect was not nullified by reverse frequency dependence on pacing. "dl-Sotalol is highly effective in the suppression of sustained monomorphic VT, inducible by programmed stimulation" (78). In the light of these recent reports it is pertinent to recall a much earlier study (1982), which showed that treatment for one year of post-MI patients with sotalol caused a 19% reduction in mortality, in line with, but not superior to, the results of trials with other β-blockers (79).

Sotalol and Proarrhythmia

In none of the above studies was proarrhythmia reported as a serious problem, but in four earlier papers by Neuvonen et al. it was claimed that sotalol could induce arrhythmias. In the first, two patients had taken large overdoses of sotalol, one after having ingested diazepam, chlordiazepam, and alcohol; the other, barbiturate and alcohol (80). The second paper reported sotalol overdosage in 6 patients, of whom 4 had taken alcohol, and 2 diazepam as well (81). The third paper recounted how a patient, taking 400 to 600 mg sotalol daily in addition to hydrochlorothiazide and hydralazine, fainted during a flight to Leningrad, again after taking alcohol. There was no evidence that his syncope was of cardiac origin but this was assumed (82). In the fourth paper, association of QT$_C$ prolongation related to sotalol concentration was observed in thirty-three hypertensive patients, but no arrhythmias were reported. Nevertheless, a warning was given that "the probability of sotalol's own arrhythmogenic action, in association with the prolonged QT$_C$, should be kept in mind" (83).

d-Sotalol

A comparison of the effects of d- and dl-sotalol in anesthetized dogs, showed that, in plasma concentrations, pharmacokinetics, effects in decreasing heart rate, lowering BP, prolonging ventricular ERP, and PR, QT and QT_C intervals, the two drugs were identical, neither of them altering QRS (84). The drugs were also similar on administration to conscious dogs. The results suggest that the Class 3 action predominated, and that unless there is a substantial background sympathetic activity, d- and dl-sotalol are indistinguishable. Similarly, in isolated preparations of the rabbit AV node, microelectrode recording revealed that d- and l-sotalol had identical electrophysiological effects. Maximum diastolic potential, action potential amplitude and V_{max} were unchanged. Spontaneous frequency was reduced and AV conduction delayed by APD prolongation to the same extent by both drugs (85). In humans, in spite of a lack of β_1-receptor blockade, confirmed by absence of occupancy of β_1-adrenoceptors in an *in vitro* radio receptor assay on plasma samples, d-sotalol reduced exercise heart rate by 7.7% ± 3.8%. "The effect is most likely a result of prolongation of the sinus node action potential duration" (86).

THE SURVIVAL WITH ORAL D-SOTALOL TRIAL (SWORD)

This trial was undertaken to test the hypothesis that a "pure" Class 3 antiarrhythmic action might improve prognosis for post-MI patients. The subjects were survivors of acute MI, with ejection fractions ≤40%, started on 200 to 400 mg d-sotalol or placebo daily, 6 to 42 days after the infarction. Also included were patients with congestive heart failure of severity II or III on the New York Heart Association scale, consequent to remote MI. The trial was terminated because the mortality was 4.6% in the d-sotalol-treated group, versus 2.7% in the placebo group ($p = .005$), the trends diverging continuously (87). As a result, the manufacturer has terminated further development of d-sotalol. Commenting on the outcome, Cobbe observed "the demise of d-sotalol puts in doubt the further development of other selective Class III antiarrhyth-

mic drugs. The highly selective dofetilide was shown to suppress inducible ventricular tachyarrhythmias and is currently the subject of a large scale mortality trial, the DIAMOND study in Denmark" (88).

NEW CLASS 3 ANTIARRHYTHMIC AGENTS

Bretylium and n-acetylprocainamide (NAPA) had been shown to possess Class 3 action, but their other properties made them less attractive than sotalol as parent compounds for the development of improved drugs. Much time and talent invested in such development may be jettisoned in the light of the SWORD preliminary results, but some of the new Class 3 agents may still be better than the old.

METHANE SULFONAMIDES

SR 33589

The unwanted effects of amiodarone were attributed, at least in part, to the presence of iodine, and Labaz synthesized a number of iodine-free derivatives, of which L 8040 and L 8462 were shown to have Class 3 activity, but were not superior to amiodarone itself (89). A more recent derivative, SR 33589 (Labaz having been meanwhile absorbed by Sanofi), incorporates the methylsulfonamide group. Like amiodarone, in addition to its Class 3 action, it exerts a noncompetitive antagonism to sympathetic effects (in rats), but does not alter T_3, T_4, or rT_3 concentrations in plasma (90). It protected rats against ischemia-reperfusion-induced arrhythmias (91,92). In dogs a comparison of SR 33589 and amiodarone showed the drugs to be very similar, with slow onset (90 minutes) of bradycardia, delayed AV conduction, prolonged ventricular ERP (amiodarone > SR33589) and QT interval (93). The main difference was a reverse frequency dependence of the effect of SR 33589 on ventricular ERP.

Compound II

Connors et al. synthesized twenty-two 4-methanesulfonamido-phenoxypropanolamines, which were tested for β-blocking activity, prolongation of APD and inhibition of I_K, elucidating structure-action relations for these effects (94). Some of the compounds were among the most potent agents in prolonging the cardiac action potential described to date, and compound II was selected for further study. It caused bradycardia and prolonged APD by blocking I_K (95).

Dofetilide

UK 68798 (dofetilide) has a methylsulfonamide group at each end, and is at least 100 times more potent than d-sotalol in prolonging ERP and QT interval in dogs, and has a dose-dependent positive inotropic action (96). UK 68798 increased ERP by 24 ms ± 10 ms at a pacing cycle length of 700 ms, and by 20+ 12 ms at CL 250 ms. In dogs with a previous MI, induction of VT could be prevented by a lengthening of ERP in the epicardial region surrounding the infarct. In contrast, if VF was induced, it was dependent on a rapid reentrant or focal mechanism that was not altered by UK 68798 (97). This may be explained by the observation that significant increases in ERP were not demonstrated for dofetilide and E-4031 in simultaneous conditions of increased $[K+]_o$ and rapid pacing (98). The reverse frequency dependence of dofetilide was confirmed.

In isolated guinea pig right ventricular papillary muscles, dofetilide attenuated the APD shortening and negative inotropic effects of hypoxia and of the K-channel-opener nicorandil (99). In dogs in which heart failure had been induced by microembolism of the area supplied by the left main coronary artery, dofetilide neither attenuated nor exacerbated the depression of ventricular function, but produced bradycardia and prolonged ERP and QT interval (100). In isolated guinea pig ventricular myocytes, dofetilide blocked I_K, preferentially its rapid component, and the effect

on ERP correlated well with its binding to the K channel (101).

Dofetilide has been used extensively in patients. Doses of 1.5, 3.0, and 4.5 mg.kg-1 IV in eighteen subjects with coronary artery disease increased QT by 36 ms, 52 ms, and 83 ms, respectively. The mean elimination half-life was 9.7 h (102). Electrophysiological studies of dofetilide were undertaken in eighteen patients with stable angina pectoris. Dofetilide, 3 mg.kg^{-1} + 1.5 mg.kg^{-1} maintenance or 5 mg.kg^{-1} + 2.5 mg.kg^{-1} maintenance, increased atrial ERP by 10% to 23% and ventricular ERP by 6% to 16%, but there was no effect on PA (p-wave to atrial deflection), AH (atrial to His bundle deflection), HV (His bundle to ventricular deflection), PR (p-wave to R-wave) intervals or on QRS (103). In twenty-four patients with atrial fibrillation (AF) or atrial flutter (AFL) of mean duration 32 to 43 days, 10/19 with AF and 4/5 with AFL were restored to sinus rhythm by dofetilide IV, after a mean dose-conversion interval of 4 ± 52 min (range 3 to 100 min) (104).

In patients with angina pectoris, dofetilide prolonged monophasic action potentials in atria and ventricles. This study confirmed the inappropriateness of calculating the QT_C. "The QT_C did not accurately reflect either changes in the plasma concentration of dofetilide, or the induced changes in refractory period and monophasic action potential duration" (105). In 50 patients with sustained monomorphic ventricular tachycardia, who had been unresponsive to up to seven (median, three) other drugs, intravenous dofetilide completely suppressed or slowed inducible VT in 17/41 subjects at doses of 3 to 15 µg.kg^{-1}, but the nine remaining patients, who had received only 1.5 µg.kg^{-1} dofetilide, did not respond (106).

E 4031

The distinction between the rapid and slow components of I_K was made possible by the selectivity for I_{Kr} of E 4031. Although I_{Kr} activates more rapidly, it deactivates more slowly than I_{Ks} (107). In a canine model of MI, E 4031 protected against lethal ischemic ventricular arrhythmias (108), and in another (pericarditis) model, in which atrial arrhythmias can be induced, E 4031 prolonged cycle length in atrial fibrillation and flutter and converted both to sinus rhythm (109).

In patients with supraventricular arrhythmias E 4031 prolonged ERP of the atrium. In subjects with accessory pathways the ERP of the antegrade, not the retrograde, pathway was prolonged. No Class 1 action on conduction was apparent, but the anticipated lengthening of QT interval was observed. At the doses used, sufficient to induce a 10% increase of QT interval, "we cannot confirm the clinical usefulness of E 4031 in patients with supraventricular tachyarrhythmias" (110).

Sematilide

The effects of sematilide resemble those of other methanesulfonamides, prolongation of APD by preferential block of I_{Kr}, without activity on sodium or calcium channels. It exhibits a much more prominent reverse frequency dependence for APD lengthening in guinea pig ventricular myocytes (111), but exerts a positive inotropic action (112). In fourteen patients with chronic high-frequency nonsustained ventricular arrhythmias, IV infusions of sematilide prolonged QT interval in a dose-related manner, without change of PR interval or QRS. Heart rate was reduced at high concentrations (>2μg.ml^{-1}) by which QT$_C$ was prolonged by 25%. The mean elimination half-life was 3.6 ± 8h. Plasma concentrations >0.8μg.ml^{-1} suppressed arrhythmias in five subjects, but exacerbated them in three others, one of whom required cardioversion (113).

In an analysis of QT dispersion from minimum and maximum QT intervals in twelve ECG leads, sematilide was compared with amiodarone and sotalol in patients with heart disease and cardiac arrhythmias, twenty-six patients for each drug. All three drugs significantly prolonged QT (and QT$_C$), but only amiodarone significantly reduced dispersion.

Amiodarone and sotalol (both with additional Class 2 action) lowered heart rate, but sematilide did not (114). In patients with ventricular arrhythmias and sustained VT inducible at EPS, oral sematilide (133 ± 9 mg t.i.d.) prolonged QT interval, without change of PR, HV, or QRS. ERP was significantly increased on pacing at a cycle length of 600 ms, but the effect was reduced as frequency increased, disappearing at CL 300 ms. Sematilide suppressed the induction of sustained VT in 41% of patients (115).

WAY-123,398

In contrast to sematilide, another methanesulfonamide Class 3 agent, WAY 123,398, did not exhibit reverse frequency dependence of the effect on APD. It was, however, thirteen-fold less potent than E 4031 and thirty-fold less than dofetilide. In anesthetized dogs there was considerable selectivity for prolongation of atrial over ventricular APD. In most respects WAY 123,398 resembled the other methanesulfonamides: bradycardia, prolonged ERP and QT, absence of Class 1 action. The threshold for induction of VF was increased (116). In feline cardiac myocytes WAY 123,398 "blocked I_K tail currents with an IC$_{50}$ of about 0.1 μM, a concentration compatible with the EC$_{50}$ for APD prolongation in dog Purkinje fibers. Block occurs rapidly (during the first pulse) and reverses very slowly (with a time-constant of several seconds)." There was no block of sodium or calcium channels, or of I_{TO} (117).

Ibutilide

In experiments on single guinea pig cardiac cells it was found that although ibutilide is a methane sulfonanilide, it not only blocked I_{K-r}, but at low concentrations (10^{-8} M) prolonged APD by increasing a late-activating inward (slow) *sodium current*, I_{Na-s}. On theoretical grounds, prolongation of APD by activation of an inward sodium current would appear unwise, entailing a risk of provoking an out-of-phase depolarization, especially if outward potassium current is simultaneously depressed. At very high concentrations (10^{-5} M) an outward current was activated, so that the concentration-effect curve on APD was bell-shaped. The time-independent outward current activated by high concentrations of ibutilide reversed the APD prolongation by the other K-channel blockers. It was not blocked by 1 μM gliburide, 5 mM 4-aminopyridine, or 5 μM E-4031, but was reduced by about 40% by 20 mM tetraethyl-ammonium (118). Ibutilide is an analogue of sotalol, but has no beta-blocking activity. In a canine model of atrial fibrillation induced by a Y-cut in the atrium, ibutilide abolished the arrhythmia in 8 out of 8 dogs and was more effective than sematilide, encainide, or lidocaine (119). In cells taken from a murine atrial tumor (AT-1 cells) ibutilide blocked I_{K-r} (EC_{50} 20 nM) as does dofetilide (EC_{50} 12 nM) (120).

Two hundred sixty-six patients with atrial fibrillation (AF, 133) or atrial flutter (AFL, 133) of duration 3 to 45 days were randomly distributed for treatment with placebo, with ibutilide 1.0 + 0.5 mg infused IV over 10 minutes, or with ibultilide 1.0. + 1.0 mg ibutilide IV. After a mean interval of 27 minutes from the infusion, 47% of the ibutilide patients were converted to sinus rhythm, there being no significant difference in the conversion rate for the lower (44%) or higher doses (49%). The conversion rate on placebo was 2%. Of the 180 patients receiving ibutilide, however, 15 developed a polymorphc ventricular tachycardia, 12 nonsustained, the other 3 requiring cardioversion (121). Ibutilide has been approved in the USA for conversion of atrial fibrillation.

CLASS 3 AGENTS WITHOUT A METHYLSUPHONAMIDE GROUP

There are now so many compounds that delay cardiac repolarization, but which have not reached the status of antiarrhythmic drugs meriting clinical evaluation, that a detailed discussion is beyond the scope of this chapter. Structure-action relations of Class 3 agents have been reviewed by Morgan and Sullivan (122).

Clofilium

Clofilium was an early example of a drug purposely designed to block potassium current, but its poor absorption on oral administration restricted its clinical utility.

Almokalant

In guinea pig ventricular myocytes almokalant inhibited the rapidly activating component of I_K (123). In anesthetized dogs almokalant 1μmol.kg^{-1} IV prolonged atrial and ventricular ERP, and the ventricular MAP and QT interval. The effect of almokalant on atrial refractoriness was very little influenced by the paced heart rate and was twice as large as the corresponding effect on the ventricle (124). In pigs, the absence of reverse frequency dependence of the effect on APD and ERP was confirmed, as was the greater activity on atria than on ventricle. There was no Class 1 action (125).

In microelectrode recordings from isolated rabbit ventricular muscle and Purkinje cells— unlike dofetilide and E-4031, which prolonged APD more in Purkinje cells than in the ventricle—when compared at concentrations giving a 15% increase of the action potential duration in ventricular muscle, almokalant was significantly less effective in the Purkinje fibers than the other drugs. Almokalant showed a less unfavorable profile in terms of dispersion of repolarization. All three compounds had a positive inotropic action (126).

In healthy human volunteers almokalant increased atrial and ventricular ERP and QT interval, but QRS and PQ interval were unchanged. Pacing at 60, 100, and 120 beats/min "did not affect the percentage of prolongation of QT." Spontaneous heart rate was slightly decreased. The half-life of the drug was short (3h) (127). Right ventricular monophasic action potentials were recorded in human volunteers at pacing frequencies of 100 and 120 beats.min^{-1}. Almokalant was injected IV as an initial bolus followed by an infusion to produce plasma concentrations in the ranges 20 to 50, 50 to 100, and 100 to 150 nM maintained for 1 hour. At a mean concentration of 116 nM MAP duration was increased by 20% at 100 beats.min^{-1} and by 19% at 120 beats.min^{-1}, confirming that almokalant does not exhibit reverse frequency dependence in man, as previously observed in animals (128).

In patients with supraventricular tachycardia (SVT) involving an accessory pathway, almokalant terminated the SVT (induced by transesophageal stimulation) in 56% of cases and prevented induction in 67% (129). In ten post-MI patients with complex ventricular arrhythmias, 4.5 mg (12.5 μmol) of almokalant or placebo was infused IV over 10 minutes. At the end of the infusion the corrected QT was 548 ± 53 ms after almokalant and 445 ± 18ms after placebo ($p = .0015$). The drug was rapidly cleared from the plasma at 11 ± 1 ml.min^{-1}.kg^{-1}, the elimination half life being 2.45h ± 0.1h (130).

Tedisamil

Unlike the methanesulfonamides, tedisamil selectively blocks I$_{TO}$ in rat and human ventricular myocytes. In human papillary muscle APD was prolonged by 3 mM tedisamil. Concentrations of 1 to 10 mM reduced peak I$_{TO}$, but current at the end of a 300 ms pulse was not affected. The inactivation time constant of I$_{TO}$ was decreased by tedisamil from 65.5 ms to 20.3 ms (131). In isolated rabbit hearts tedisamil blocked I$_{TO}$ and IK$_1$. Concentrations of 1, 3, and 10 mM increased the ventricular ERP from 120 ± 18ms to 155 ms, 171 ms, and 205 ms, respectively. Hypoxia-reperfusion induced VF in 5 out of 6 control hearts, but in 4 out of 6 in the presence of 1 mM tedisamil, and in 0 out of 6 in the presence of 3 mM. Electrically induced VF was terminated by the infusion of 0.3 ml of 10 mM tedisamil into the aortic cannula of Langendorff-perfused hearts (132).

In ten patients with coronary stenosis (>60%) LV MAPs were recorded during pacing. Tedisamil 0.3 mg.kg-1 reduced spontaneous heart rate by 12%, and increased MAP

duration by 16%, without change of QRS. The effect on MAP exhibited reverse frequency dependence. The prolongation of MAP was decreased as the pacing cycle length was reduced from 600ms to 400ms (133).

CONCLUSION

There are many other compounds with Class 3 action combined with other effects, which may or may not reach further stages of development. For example, azimilide (NE 10064) blocks I$_K$ and calcium current (134). MS-551 blocks not only I$_K$ but I$_{TO}$ and I$_{K1}$ as well, but whether this is advantageous is unknown (135). The failure of d-sotalol to protect post-MI patients does not indicate that pure Class 3 agents are valueless. They may prove useful for preventing or terminating reentrant arrhythmias, either alone or in combination with other drugs.

Since drugs with Class 2 action have long been proven to reduce mortality in post-MI patients, perhaps, in a search for improvement, a new remedy (or placebo) should be administered in addition to a Class 2 drug, which would be given to both groups.

REFERENCES

1. Durrer D, van Dam RTh, Freud GE, Janse MJ, Meijler FL, Arzbaecher RC. Total excitation of the isolated human heart. *Circulation* 1970; 41:899–912.
2. Davidson NC, Darbar D, Luck J, Main G, Pringle TH, McNeill GP, Struthers AD. QT interval dispersion predicts cardiac death in patients with peripheral vascular disease. *Br Heart J* 1995; 73:(Suppl 3): 54.
3. Yuan S, Wohlfart B, Olsson SB, Blomström-Lundquist C. The dispersion of repolarization in patients with ventricular tachycardia. *Eur Heart J* 1995; 16:68–76.
4. Yuan S, Blomström-Lundquist C, Pehrson S, Pripp C-M, Wohlfart B, Olsson SB. Dispersion of repolarization following double and triple programmed stimulation. *Eur Heart J* 1996; 16:1080–1091.
5. McLaughlin NB, Campbell RWF, Murray A. Comparison of automatic QT measurement techniques in the normal 12-lead electrocardiogram. *Br Heart J* 1995; 74:84–89.
6. Sanguinetti MC, Jurkiewicz NK. Two components of cardiac delayed rectifier. *J Gen Physiol* 1990; 96: 195–215.
7. Li G-R, Feng J, Yue L, Carrier M, Nattel S. Evidence for two components of delayed rectifier K$^+$ current in human ventricular myocytes. *Circ Res* 1995; 78: 689–696.

8. Daleau P, Turgeon J. Triamterine inhibits the delayed rectifier potassium current (I_K) in guinea pig ventricular myocytes. *Circ Res* 1994; 74:1114–1120.

9. Veldkamp MW, van Ginneken ACG, Opthof T, Bouman LN. Delayed rectifier channels in human ventricular myocytes. *Circulation* 1995; 92:3447–3504.

10. Yue L, Feng J, Li G-R, Nattel S. Characterization of an ultrarapid delayed rectifier potassium current involved in canine atrial repolarization. *J Physiol* 1996; 496:647–662.

11. Rees SA, Curtis MJ. Specific I_{K1} blockade: a new antiarrhythmic mechanism? *Circulation* 1993; 87:1979–1989.

12. Carmeliet E. Chloride ions and the membrane potential of Purkinje fibres. *J Physiol* 1961; 156:375–388.

13. Dudel J, Peper K, Rudel R, Trautwein W. The dynamic chloride component of membrane current in Purkinje fibres. *Pflügers Arch ges Physiol* 1967; 295:197—212.

14. Fozzard HA, Hiraoka M. The positive dynamic current and its inactivation in cardiac Purkinje fibres. *J Physiol* 1973; 234:569–586.

15. McAllister RE, Noble D, Tsien RW. Reconstruction of the electrical activity of cardiac Purkinje fibres. *J Physiol* 1975; 251:1–60.

16. Kenyon JL, Gibbons WR. Influence of chloride, potassium and tetraethylammonium on the early outward current of sheep cardiac Purkinje fibers. *J Gen Physiol* 1979; 73:117–138.

17. Kenyon JL, Gibbons WR. 4-Aminopyridine and the early outward current of sheep cardiac Purkinje fibers. *J Gen Physiol* 1979; 73:139–157.

18. Carmeliet E, Verdonck F. Reduction of potassium permeability by chloride substitution in cardiac cells. *J Physiol* 1977; 265:193–206.

19. Ten Eick RE, Whalley OW, Rasmussen HH. Connections: heart disease, cellular electrophysiology and ion channels. *FASEB J* 1992; 6:2568–2580.

20. Shimoni Y, Severson D, Giles W. Thyroid states and diabetes modulate regional differences in potassium currents in rat ventricle. J Physiol 1995; 488:673–688.

21. Kawano S, Hirayama Y, Hiraoka M. Activation mechanism of Ca^{2+}-sensitive transient outward current in rabbit ventricular myocytes. *J Physiol* 1995; 486:593–604.

22. Amos GJ, Wettwer E, Metzger F, Li Q, Himmel HM, Ravens U. Differences between outward currents of human atrial and subepicardial ventricular myocytes. *J Physiol* 1996; 490:31–50.

23. Mori K, Saito T, Masuda Y, Nakaya H. Effects of Class III antiarrhythmic drugs on the Na^+-activated K^+ channels in guinea pig ventricular cells. *Br J Pharmacol* 1996; 119:133–141.

24. Allen DG, Morris PG, Orchard CH, Pirolo JS. A nuclear magnetic resonance study of metabolism in the ferret heart during hypoxia and inhibition of glycolysis. *J Physiol* 1985; 361:185–204.

25. Cowan JC, Vaughan Williams EM. The effects of various fatty acids on action potential shortening during sequential periods of ischaemia and reperfusion. *J Mol Cell Cardiol* 1980; 12:347–369.

26. Vleugels A, Carmeliet E. Hypoxia increases potassium efflux from mammalian myocardium. *Experientia* 1976; 32:483–484.

27. Noma A, Tohru S. Properties of adenosine-triphosphate-regulated potassium channels in guinea pig ventricular cells. *J Physiol* 1985; 363:463–480.

28. Thuringer D, Cavero I, Coraboeuf E. Time-dependent fading of the activation of K_{ATP} channels induced by aprikalim and nucleotides, in excised membrane patches from cardiac myocytes. *Br J Pharmacol* 1995; 115:117–127.

29. Findlay I. The ATP-sensitive potassium channel of cardiac muscle and action potential shortening during metabolic stress. *Cardiovasc Res* 1994; 28:760–761.

30. Nakamura TY, Faivre J-F, Coetzee WA. Spontaneous and thiol-group-induced changes in ATP-sensitivity of K_{ATP} channels from isolated guinea pig ventricular myocytes. *J Physiol* 1995; 483P:10P.

31. Li G-R, Feng J, Shrier A, Nattel S. Contribution of ATP-sensitive potassium channels to the electrophysiological effects of adenosine in guinea pig atrial cells. *J Physiol* 1995; 484:629–643.

32. von Kügelgen I, Stoffel D, Starke K. P_2-purinoceptor-mediated inhibition of noradrenaline release in rat atria. *Br J Pharmacol* 1995; 115:247–254.

33. Escande D, Henry P. Potassium channels as pharmacological targets in cardiovascular medicine. *Eur Heart J* 1993; 14:(Suppl B) 2–9.

34. Nichols CG, Makhina EN, Pearson WL, Sha Q, Lopatin AN. Inward rectification and implications for cardiac excitability. *Circ Res* 1996; 78:1–7.

35. Roberds SL, Knoth KM, Po S, Blair TA, Bennett PB, Hartshorne RP, Snyders DJ, Tamkun MM. Molecular biology of the voltage-gated potassium channels of the cardiovascular system. *J Cardiovasc Electrophysiol* 1993; 4:68–80.

36. Charlier R. A new antagonist of adrenergic excitation not producing competitive receptor blockade. *Br J Pharmacol* 1970; 39:668–674.

37. Calkins H, Sousa J, El-Atassi R, Schmaltz S, Kadish A, Morady F. Reversal of antiarrhythmic drug effects by epinephrine. Quinidine versus amiodarone. *J Am Coll Cardiol* 1992; 19:347–352.

38. Bacq ZM, Blakeley ACH, Summers RJ. The effects of amiodarone, an α- and β-receptor antagonist, on adrenergic transmission in the cat spleen. *Biochem Pharmacol* 1976; 25:1195–1199.

39. Twidale N, Roberts-Thomson P, McRitchie RJ. Comparative hemodynamic effects of amiodarone, sotalol and d-sotalol. *Am Heart J* 1993; 126:122–129.

40. Sato R, Koumi S-I, Singer DH, Hisatome I, Jia H, Eager S, Wasserstrom JA. Amiodarone blocks the inward rectifier potassium channel in isolated guinea pig ventricular cells. *J Pharmacol Exp Ther* 1994; 269:1213–1219.

41. Varró A, Virág L, Papp JGy. Comparison of the chronic and acute effects of amiodarone on the calcium and potassium currents in rabbit isolated cardiac myocytes. *Br J Pharmacol* 1996; 117:1181–1186.

42. Balser JR, Bennett PB, Hondeghem LM, Roden DM. Suppression of time-dependent outward current in guinea pig ventricular myocytes. *Circ Res* 1991; 69:519–529.

43. Huikiri HV, Yli-Mä YRY. Frequency dependent effects of d-sotalol and amiodarone on the action potential duration of the human right ventricle. *Pace* 1992; 15:2103–2107.

44. Mason JW, Hondeghem LM, Katzung BG. Amiodarone blocks inactivated cardiac sodium channels. *Pflügers Arch* 1983; 396:39–81

45. Chiamvimonvat N, Mitchell LB, Gillis AM, Wyse DG, Sheldon RS, Duff HJ. Use-dependent electrophysio-

logical effects of amiodarone in coronary artery disease and inducible ventricular tachycardia. *Amer J Cardiol* 1992; 70:598–604

46. Chen F, Wetzel GT, Klitzner TS. Acute effects of amiodarone on sodium currents in isolated neonatal ventricular myocytes: comparison with procainamide. *Dev Pharmacol Ther* 1992; 19:118–130

47. Vaughan Williams EM. The classification of antiarrhythmic drugs reviewed after a decade. In the AN Richards Symposium. Mechanism and treatment of cardiac arrhythmias. Ed HJ Reiser, LN Horowitz. *Urban-Schwartzenberg*, Baltimore. 1985. 153–161.

48. Quinteiro RA, Biagetti MO, de Forteza E. Relationship between V_{max} and conduction velocity in uniform anisotropic canine ventricular muscle: differences between the effects of lidocaine and amiodarone. *J Cardiovasc Pharmacol* 1990; 16: 931–939

49. Quinteiro RA, Biagetti MO. Chronic versus acute effects of amiodarone on the V_{max} conduction velocity relationship and on the space constant in canine myocardium. *J Cardiovasc Pharmacol* 1994; 24: 122–132.

50. Harris L, Roncucci R. *Amiodarone*. Paris: Médecine et Sciences Internationales; 1986.

51. Gøtzsche LB-H. β-adrenergic receptors, voltage-operated Ca^2-channels, nuclear triiodothyronine receptors and triiodothyronine concentration in pig myocardium after long-term low-dose amiodarone treatment. *Acta Endocrinologica* 1993; 129:337–347.

52. Awaji T, Wu ZJ, Hashimoto K. Acute antiarrhythmic effects of intravenously administered amiodarone on canine ventricular arrhythmias. *J Cardiovasc Pharmacol* 1995; 26:869–878.

53. Hou ZY, Chang M-S, Chen C-Y, Tu M-S, Lin S-L, Chiang H-T, Woosley RL. Acute treatment of recent-onset atrial fibrillation and flutter with a tailored dosing regimen of intravenous amiodarone. A randomized, digoxin-controlled study. *Eur Heart J* 1995; 16:521–528.

54. Camm AJ, Julian DG, Janse G, Munuz A, Schwartz P, Frangin G. The European Myocardial Infarct Amiodarone Trial (EMIAT). *Am J Cardiol* 1993; 72: 95F–98F.

55. Burkart F, Pfisterer M, Kiowski W, Follath F, Burckhardt D. Effect of antiarrhythmic therapy on mortality in survivors of myocardial infarction with asymptomatic complex ventricular arrhythmias: Basel Antiarrhythmic Study of Infarct Survival (BASIS). *J Am Coll Cardiol* 1990; 16:1711–1718.

56. Pfisterer M, Kiowski W, Burckhardt D, Follath F, Burkart F. Beneficial effect of amiodarone on cardiac mortality in patients with asymptomatic complex ventricular arrhythmias after acute myocardial infarction and preserved but not impaired left ventricular function. *Am J Cardiol* 1992; 69:1399–1402.

57. Cairns JA, Connolly SJ, Gent M, Roberts R. Post-myocardial infarction mortality in patients with ventricular premature depolarizations. Canadian Amiodarone Myocardial Infarction Arrhythmia Trial (CAMIAT). Pilot Study. *Circulation* 1991; 84: 550–557.

58. Ceremuzynski L, Kleczar E, Krzeminska M, Kulh J, Nartowitz E, Smielak-Korombel J, Dyduszynski A, Maciejewicz J, Zaleska T, Lazaczyk-Kedzia E, Motyka J, Paczkowska B, Sczaniecka O, Yusuf S. Effect of amiodarone on mortality after myocardial infarction: a double-blind, placebo-controlled, pilot study. *J Am Coll Cardiol* 1992; 20:1056–1062.

59. Weinberg BA, Miles WM, Klein LS, Bolander JE, Dusman RE, Stanton MS, Heger JJ, Langefeld C, Zipes DP. Five-year follow-up of 589 patients treated with amiodarone. *Am Heart J* 1993; 125:109–120.

60. Pfisterer ME, Kiowski W, Brunner H, Burckhardt D, Burkart F. Long-term benefit of 1-year amiodarone treatment for persistent complex ventricular arrhythmias after myocardial infarction. *Circulation* 1993; 87:309–311.

61. Podrid PJ. Therapy with and assessment of Class III antiarrhythmic agents in different patient populations. *J Cardiovasc Pharmacol* 1992; 20:(Suppl 2) S44–S58.

62. Hohnloser SH, Zabel M, Zehender M, Meinertz T, Just H. Efficacy of intravenously administered amiodarone for short-term control of serious arrhythmias. *J Cardiovasc Pharmacol* 1992; 20:(Suppl 2) S63–S69.

63. Kim SG, Mannino MM, Chou R, Roth S, Roth JA, Ferrick KJ, Fisher JD. Rapid suppression of spontaneous ventricular arrhythmias during oral amiodarone loading. *Ann Intern Med* 1992; 117:197–201.

64. Cobbe SM, Manley BS. Cellular electrophysiology of amiodarone in cardiac ischaemia. *Br J Clin Pract* 1986; 40(Suppl 44):104.

65. Mayuga RD, Singer DH. Effects of intravenous amiodarone on electrical dispersion in normal and ischaemic tissues, and on arrhythmia inducibility: monophasic action potential studies. *Cardiovasc Res* 1992; 26:571–579.

66. Follath F. The utility of serum drug level monitoring during therapy with Class III antiarrhythmic agents. *J Cardiovasc Pharmacol* 1992; 20(Suppl 2):S41–S43.

67. Tisdale JE, Follin SL, Ordelova A, Webb CR. Risk factors for the development of specific adverse effects associated with amiodarone. *J Clin Pharmacol* 1995; 35:351–356.

68. Nathan AW, Hellestrand KJ, Bexton RS, Ward DE, Spurrell RAJ, Camm AJ. Electrophysiological effects of sotalol—just another beta-blocker? *Br Heart J* 1982; 47:515–520.

69. Bennett DH. Acute prolongation of myocardial refractoriness by sotalol. *Br Heart J* 1982; 47:521–526.

70. Xu H, Villafane J, McCormack J, Stolfi A, Gelband H, Pickoff AS. Comparison of the electrophysiological effects of intravenous sotalol and propranolol on the immature mammalian heart. *J Cardiovasc Pharmacol* 1989; 13:925–929.

71. Varró A, Nánási PP, Lathrop DA. Effect of sotalol on transmembrane ionic currents responsible for repolarization in cardiac ventricular myocytes from rabbit and guinea pig. *Life Sci* 1991; 49:7–12.

72. Schmitt C, Beyer T, Karch M, Montero M, Hilbel T, Brachmann J, Kübler W. Sotalol exhibits reverse use-dependent action on monophasic action potentials in normal but not in infarcted canine ventricular myocardium. *J Cardiovasc Pharmacol* 1992; 19:487–492.

73. Wang J, Bourne GW, Wang Z, Villamaire C, Talajic M, Nattel S. Comparative mechanisms of antiarrhythmic drug action in experimental atrial fibrillation. *Circulation* 1993; 88:1030–1044.

74. Anastasiou-Nana MI, Gilbert EM, Miller RH, Singh S, Freedman RA, Keefe DL, Saksena S, MacNeil DJ, Anderson JL. Usefulness of d,l sotalol for suppression

of chronic ventricular arrhythmias. *Am J Cardiol* 1991; 67:511–516.

75. Kus T, Campa MA, Nadeau R, Dubul M, Kaltenbrunner W, Shenasa M. Efficacy and electrophysiologic effects of oral sotalol in patients with sustained ventricular tachycardia caused by coronary artery disease. *Am Heart J* 1992; 123:82–89.

76. Reisinger J, Shenasa M, Lubinski A, Hendrick SG, Haverkamp W, Chen X, Breithardt G, Borggrefe M. Clinical implications of pleomorphic ventricular tachycardia on oral sotalol therapy. *Eur Heart J* 1995; 16:377–382.

77. Ho DSW, Zecchin RP, Cooper MJ, Richards DAB, Uther JB, Ross DL. Rapid infusion of d,l sotalol: time to onset of effects on ventricular refractoriness and safety. *Eur Heart J* 1995; 16:81–86.

78. Kühlkamp V, Mermi J, Mewk C, Braun U, Seipel L. Long-term efficacy of d,l sotalol in patients with sustained ventricular tachycardia refractory to Class 1 antiarrhythmic drugs. *Eur Heart J* 1995; 16:1625–1631.

79. Julian DG, Prescott RJ, Jackson FS, Szekely P. Controlled trial of sotalol for one year after myocardial infarction. *Lancet* 1982; (1):1142–1147.

80. Neuvonen PJ, Elonen E, Tarssanen L. Sotalol intoxication. Two patients with concentration-effect relationships. *Acta Pharmacol Toxicol* 1979; 45:53–57.

81. Neuvonen PJ, Elonen E, Vuorenmaa T, Laakso M. Prolonged QT interval and severe tachyarrhythmias. Common features of sotalol intoxication. *Eur J Clin Pharmacol* 1981; 20:85–89.

82. Laakso M, Pentkäinen PJ, Pyörälä K, Neuvonen PJ. Prolongation of the QT interval caused by sotalol— possible association with ventricular tachyarrhythmias. *Eur Heart J* 1981; 2:353–358.

83. Neuvonen PJ, Elonen E, Tanskanen A, Tuomilehto J. Sotalol prolongation of the QT_C interval in hypertensive patients. *Clin Pharmacol Ther* 1982; 32:25–32.

84. Gomoll AW, Lekich RF, Bartek MJ, Comereski CR, Antonaccio MJ. Comparability of the electrophysiologic responses and plasma and myocardial tissue concentrations of sotalol and its d stereoisomer in dogs. *J Cardiovasc Pharmacol* 1990; 16:204–211.

85. Beyer T, Brachmann J, Kübler W. Comparative effects of d-sotalol and l-sotalol on the atrioventricular node of the rabbit heart. *J Cardiovasc Pharmacol* 1993; 22:240–246.

86. Yasuda SA, Barbey JT, Funck-Brentano C, Wellstein A, Woosley RL. d-Sotalol reduces heart rate in vivo through a β-adrenergic-receptor-independent mechanism. *Clin Pharmacol Ther* 1993; 53:426–432.

87. Waldo AL, Camm AJ, de Ruyter H, Friedman PL, MacNeil DJ, Pitt B. Preliminary mortality results from the SURVIVAL WITH ORAL D-SOTALOL TRIAL (SWORD). *J Am Coll Cardiol* 1995; 25:(Suppl)15A.

88. Cobbe SM. Class III antiarrhythmics: put to the SWORD? Heart 1996; 75:111–113.

89. Vaughan Williams EM, Polster P. The effect on cardiac muscle of two drugs related to amiodarone, L 8040 and L 8462. *Eur J Pharmacol* 1974; 25:241–247.

90. Chatelan P, Meysmans L, Mattéazzi JR, Beaufort P, Clinet M. Interaction of the antiarrhythmic agents SR 33589 and amiodarone with β-adrenoceptors and adenylate cyclase in rat heart. *Br J Pharmacol* 1995; 116:1949–1952.

91. Bruyninckx C, Ramboux J, Chatelain P, Manning A. SR 33589, a new amiodarone-like agent: effect on reperfusion-induced arrhythmias in anaesthetized rats. *J Mol Cell Cardiol* 1992; 24 (Suppl V):S68.

92. Manning AS, Bruyninckx C, Ramboux J, Chatelain P. SR 33589, a new amiodarone-like agent: effect on ischemia and reperfusion-induced arrhythmias in anaesthetized rats. *J Cardiovasc Pharmacol* 1995; 26:453–461.

93. Manning A, Thisse V, Hodeige D, Richard J, Heyndrickx JP, Chatelain P. SR 33589, a new amiodarone-like antiarrhythmic agent: electrophysiologic effects in anesthetized dogs. *J Cardiovasc Pharmacol* 1995; 25:252–261.

94. Connors SP, Dennis PD, Gill EW. The synthesis and potassium channel blocking activity of some (four-methane-sulfonamidophenoxy) propanolamines as potential Class III antiarrhythmic agents. *J Med Chem* 1991; 34:1570–1577.

95. Lei M, Brown H, Terrar D. Effects of Compound II, a novel sotalol analogue, on delayed rectifier current, I_K, and spontaneous activity in isolated sinoatrial node cells of the rabbit. *J Physiol* 1996; 491P:160P.

96. Wallace AA, Stupienski RF, Brookes LM, Selnick HG, Clarenson DA, Lynch JL. Cardiac electrophysiologic and inotropic actions of new and potent methanesulfonamide Class III antiarrhythmic agents in anesthetized dogs. *J Cardiovasc Pharmacol* 1991; 18:687–695.

97. Zuanetti G, Corr PB. Antiarrhythmic effects of a new Class III agent, UK-68798, during chronic myocardial infarction: evaluation using three-dimensional mapping. *J Pharmacol Exp Ther* 1991; 256:324–334.

98. Baskin Eof amiodarone, sematilide and sotalol on QT dispersion. *Am J Cardiol* 1994; 74:896–900.

99. Yang T, Tande PM, Lathrop DA, Refsum H. Class III antiarrhythmic action by potassium channel blockade: dofetilide attenuates hypoxia-induced electromechanical changes. *Cardiovasc Res* 1992; 261:1109–1115.

100. Mortensen E, Yang T, Refsum H. Class III antiarrhythmic action: effects of dofetilide in acute ischemic heart failure in dogs. *J Cardiovasc Pharmacol* 1992; 19:216–221.

101. Lynch JJ, Baskin EP, Nott EM, Guinosso PJ, Hamill T, Salata JJ, Woods CM. Comparison of binding to rapidly activating delayed rectifier K+ channel, I_{Kr}, and effects on myocardial refractoriness for Class III antiarrhythmic agents. *J Cardiovasc Pharmacol* 1995; 25:336–340.

102. Sedgwick M, Rasmussen HS, Walker D, Cobbe SM. Pharmacokinetic and pharmacodynamic effects of UK-68,798, a new potential Class III antiarrhythmic drug. *Br J Clin Pharmacol* 1991; 31:515–519.

103. Sedgwick ML, Rasmussen HS, Cobbe SM. Clinical and electrophysiologic effects of intravenous dofetilide (UK-68,798), a new Class III antiarrhythmic drug, in patients with angina pectoris. *Am J Cardiol* 1992; 69:513–517.

104. Maarten JS, Pulak PE, van't Hof A, Rasmussen HS, Dunselman PH, Kingsma JH. Efficacy and safety of a new selective Class III antiarrhythmic agent dofetilide in paroxysmal atrial fibrillation and atrial flutter. *J Am Coll Cardiol* 1992; 69:417–419.

105. Sedgwick ML, Dalrymple I, Rae AP, Cobbe SM. Effects of the new Class III antiarrhythmic drug

dofetilide on the atrial and ventricular monophasic action potential in patients with angina pectoris. *Eur Heart J* 1995; 16:1641–1646.

106. Bashir Y, Thomsen P-EB, Kingsma H, Miller M, Wong C, Cobbe SM, Jordaens L, Campbell RWF, Rasmussen HS, Camm AJ. Electrophysiologic profile and efficacy of intravenous dofetilide (UK-68,798) a new Class III antiarrhythmic drug, in patients with sustained monomorphic ventricular tachycardia. *Am J Cardiol* 1995; 76:1040–1044.

107. Heath B, Terrar DA. Influence of E4031 on the deactivation of delayed rectifier potassium currents in guinea pig isolated ventricular myocytes. *Br J Pharmacol* 1995; 116:26P.

108. Lynch JJ, Heaney LA, Wallace AA, Gehret JR, Selnick HG, Stein RB. Suppression of lethal ischemic ventricular arrhythmias by the Class III agent E4031 in a canine model of previous myocardial infarction. *J Cardiovasc Pharmacol* 1990; 15:764–775.

109. Selaki K, Murakawa Y, Inoue H, Nakajima T, Usui M, Yamashita T, Ajiki K, Oikawa N, Iwasawa K, Omata M. Effect of isoproterenol on facilitation of electrical defibrillation by E4031. *J Cardiovasc Pharmacol* 1995; 25:393–396.

110. Fujiki A, Tani M, Mizumaki K, Shimono M, Inoue H. Electrophysiologic effects of intravenous E4031, a novel Class III antiarrhythmic agent, in patients with supraventricular tachyarrhythmias. *J Cardiovasc Pharmacol* 1994; 23:374–378.

111. Sawanobori T, Adaniya H, Namiki T, Hiraoka M. Rate-dependent effects of sematilide on action potential duration in isolated guinea pig ventricular myocytes. *J Pharmacol Exp Ther* 1994; 271:302–310.

112. Yamada A, Motomura S, Hashimoto K. Comparison of direct negative chronotropic and positive inotropic effects of sematilide to those of E4031 and MS551, and the reverse frequency dependent prolongation of cardiac refractoriness of sematilide. *J Cardiovasc Pharmacol* 1996; 27:159–166.

113. Wong W, Pavlou HN, Birgersdotter UM, Hilleman DE, Nohiudden SM, Roden DM. Pharmacology of the Class III antiarrhythmic agent sematilide in patients with arrhythmias. *Am J Cardiol* 1992; 69:206–212.

114. Cui G, Sen L, Sager P, Uppal P, Singh BN. Effects of amiodarone, sematilide and sotalol on QT dispersion. *Amer J Cardiol* 1994; 74:896–900.

115. Sager PT, Nademanee K, Antimisiaris M, Pacifico A, Pruitt C, Godfrey R, Singh BN. Antiarrhythmic effects of selective prolongation of refractoriness. *Circulation* 1993; 88:1072–1082.

116. Spinelli W, Parsons RW, Colatsky TJ. Effects of WAY-123,398, a new Class III antiarrhythmic agent, on cardiac refractoriness and ventricular fibrillation threshold in anaestheticized dogs: a comparison with UK-68798, E4031 and dl sotalol. *J Cardiovasc Pharmacol* 1992; 20:913–922.

117. Spinelli W, Moubarak IF, Parsons RW, Colatsky TJ. Cellular electrophysiology of WAY-123,398, a new Class III antiarrhythmic agent: specificity of I_K block and lack of reverse use-dependence in cat ventricular myocytes. *Cardiovasc Res* 1993; 27:1580–1591.

118. Lee KS, Tsai TD, Lee EW. Membrane activity of Class III antiarrhythmic compounds: a comparison between ibutilide, d-sotalol, E-4031, sematilide and dofetilide. *Eur J Pharmacol* 1993; 234:43–53.

119. Buchanan LV, Kabell G, Gibson JK. Acute intravenous conversion of canine atrial flutter: comparison of antiarrhythmic drugs. *J Cardiovasc Pharmacol* 1995; 25:539–544.

120. Yang T, Snyders DJ, Roden DM. Ibutilide, a methane-sulfonanilide antiarrhythmic, is a potent blocker of the rapidly activating delayed rectifier K^+ current (I_{K-r}) in AT-1 cells. *Circulation* 1995; 91:1799–1806.

121. Stambler BS, Wood MA, Ellenbogen KA, Perrry KT, Wakefield LK, Vanderlugt JT and Ibutilide repeat dose study investigators. Efficacy and safety of repeated intravenous doses of ibutilide for rapid conversion of atrial flutter or fibrillation. *Circulation* 1996; 94:1613–1621.

122. Morgan TK, Sullivan ME. An overview of Class III electrophysiological agents: a new generation of antiarrhythmic therapy. *Prog Med Chem* 1992: 29:65–108.

123. Wettwer E, Grundke M, Ravens U. Differential effects of the new Class III antiarrhythmic agents almokalant, E-4031 and d-sotalol, and of quinidine, on delayed rectifier currents in guinea pig ventricular myocytes. *Cardiovasc Res* 1992; 26:1145–1152.

124. Duker G, Almgren O, Carlsson L. Electrophysiologic and hemodynamic effects of H 234/09 (almokalant), quinidine and (+)-sotalol in the anesthetized dog. *J Cardiovasc Pharmacol* 1992; 20:458–465.

125. Wiesfeld ACP, Langen CDJ, Crijns HJGM, Bel HB, Hillege HL, Wesseling H, Lie KI. Rate-dependent effects of the Class III antiarrhythmic drug almokalant on refractoriness in the pig. *J Cardiovasc Pharmacol* 1996; 27:594–600.

126. Abrahamsson C, Duker G, Lundberg C, Carlsson L. Electrophysiological effects of H 234/09 (almokalant) in vitro: a comparison with two other novel I_K blocking drugs, UK-68,798 (dofetilide) and E-4031. *Cardiovasc Res* 1993; 27:861–867.

127. Darpö B, Almgren O, Bergstrand R, Bäärnhielm C, Gottridson C, Sandstedt B, Evaardsson N. Tolerance and effects of almokalant, a new selective I_K blocking agent, on ventricular repolarization and on sino-atrial and atrio-ventricular nodal function in the heart: a study in healthy male volunteers, utilizing transoesophageal atrial stimulation. *J Cardiovasc Pharmacol* 1995; 25:681–690.

128. Darpö B, Vallin H, Almgren O, Bergstrand R, Insulander P, Edvardsson N. Selective I_K blocker almokalant exhibits Class III-specific effects on the repolarization and refractoriness of the human heart: a study of healthy volunteers using right ventricular monophasic action potential recordings. *J Cardiovasc Pharmacol* 1995; 26:530–540.

129. Darpö B, Edvaardsson N. Effect of almokalant, a selective potassium channel blocker, on the termination and inducibility of paroxysmal ventricular tachycardias: a study in patients with Wolff-Parkinson-White syndrome and atrioventricular nodal re-entrant tachycardia. *J Cardiovasc Pharmacol* 1995; 26: 198–206.

130. Wiesfeld ACP, Crijns JGM, Tobé TJM, Almgren O, Bergstrand RH, Aberg J, Haarksma J, Lie KI. Electrophysiologic effects and pharmacokinetics of almokalant, a new Class III antiarrhythmic, in patients with healed or healing myocardial infarcts and complex ventricular arrhythmias. *Am J Cardiol* 1993; 70: 990–996.

131. Ravens U, Himmel HM, Li Q, Amos GS, Wettwer E, Posival H. Is Class III antiarrhythmic action of tedisamil due to open channel block of I_{TO} in human ventricular myocytes? *Br J Pharmacol* 1995; 114:16P.

132. Chi L, Park JL, Friedrichs GS, Banglawala YA, Perez MA, Tanhehco EJ, Lucchesi BR. Effects of tedisamil (KC-8857) on cardiac electrophysiology and ventricular fibrillation in the rabbit isolated heart. *Br J Pharmacol* 1996; 117:1261--1269.

133. Bargheer K, Bode F, Klein HU, Trappe HJ, Franz MR, Lichtlen PR. Prolongation of monophasic action potential duration and the refractory period in the human heart by tedisamil, a new potassium blocking agent. *Eur Heart J* 1994; 15:1409–1414.

134. Fermini B, Jurkiewicz NK, Jow B, Guinosso PJ, Baskin EP, Lynch JJ, Salata JJ. Use-dependent effects of the Class III antiarrhythmic agent NE-10064 (azimilide) on cardiac repolarization: block of delayed rectifier potassium and L-type calcium currents. *J Cardiovasc Pharmacol* 1995; 26:259–271.

135. Nakaya H, Tohse N, Takeda Y, Kanno M. Effects of MS-551, a new Class III antiarrhythmic drug, on action potential and membrane currents in rabbit ventricular myocytes. *Br J Pharmacol* 1993; 109:157–163.

5

The Long QT Syndrome and Torsade de Pointes

E. M. Vaughan Williams

An association between a prolonged QT interval and cardiac arrhythmias has long been recognized, not only in rare cases of congenital abnormality, but as a not infrequent accompaniment of ischemic heart disease. Schwartz and Wolf (1) identified prolongation of QT interval as a predictor of sudden death in patients with myocardial infarction, and this association may account for the reluctance of some cardiologists to accept that a drug that prolonged QT could be antiarrhythmic. The surface ECG records vectors representing the algebraic sum of currents from millions of cells, and the end of the T-wave is not necessarily coterminous with the cessation of ventricular activity, because oppositely polarized vectors can cancel each other.

The QT is not a measure of APD. Although a uniform prolongation of APD would be reflected as a longer QT interval, QT itself represents the time from the first ventricular depolarization to the last differential repolarization in the axis of the recording lead, and could conceal within it a number of short APDs juxtaposed to long ones (2).

QT can be prolonged in at least four different ways:

1. Uniformly delayed conduction (revealed as widened QRS), without any change in APD (or JT interval): This is known as a Class 1c action.
2. A nonuniform delay in conduction: Some cells repolarize later than others because they depolarize later, again without change

of APD. Such delay could be caused by conduction being forced to take a long route around an obstruction (eg, infarct scar), or by slower conduction (reduced V_{max}) through partially depolarized cells (eg, ischemic or damaged). In this case QT prolongation reflects an increased dispersion of ventricular repolarization times.

3. Heterogeneity of sympathetic innervation (eg, caused by death of sympathetic nerves enmeshed in an infarcted region): The denervated myocardial cells would have longer APDs after adaptation to their denervated state, and would not respond to increased sympathetic activity. Their innervated neighbors would have shorter (normal) APDs, which would shorten further in response to sympathetic stimulation. Thus, QT prolongation represents greater dispersion of ventricular repolarization times, increasing on exertion.
4. Uniform prolongation of APD in all ventricular cells: QT is prolonged, but ventricular repolarization times are not dispersed. This is known as a Class 3 antiarrhythmic action.

In spite of initial skepticism, the antiarrhythmic effect of a Class 3 action by many different compounds has been fully established, accompanied by QT prolongation. Increased probability of arrhythmia is associated with QT prolongation only when it reflects greater dispersion of repolarization, which can be detected by MAP recording at different sites, or by analysis of QT intervals in different ECG leads (see Chapter 4, Ref. 5). Increased dispersion of QT can occur without prolongation of APD. A local shortening of APD may accompany regional ischemia, or differential sympathetic excitation. If the shortening occurs primarily in the epicardial region, T-wave amplitude would increase; conversely, if the shortening is endocardial, the T-wave may be flattened or inverted (see Fig. 1 in Chapter 4).

THE LONG QT SYNDROME (LQTS)

In 1957 Jervell and Lange-Nielsen (3) described a syndrome of congenital deafness associated with a long QT interval and sudden death. In Italy Romano et al. (4) reported a rare type of arrhythmia in children, and the following year Ward (5) described similar cases in Ireland, which he showed to be due to an inherited condition by studying the family histories of the children involved (6). In most of the cases there was no associated deafness, but a common feature was a prolonged QT interval, or some abnormality of the T-wave, even alternating concordant and discordant T-waves (alternans). Several characteristics led to an association between the precipitation of the arrhythmia and sympathetic activity. Resting heart rate (HR) was often subnormal, with a reduced HR increase on exercise. Arrhythmias, frequently causing syncope, were initiated by stress, physical or emotional.

In a series of animal experiments Schwartz et al. (7) were able to reproduce some of the features associated with LQTS, including prolongation of QT and T-wave alternans, by differential left and right sympathetic denervation and stimulation. Unilateral sympathetic denervation or block reduced the threshold for induction of fibrillation electrically, and prolonged QT interval, as would be expected (Chapter 4) from the APD lengthening associated with sympathetic denervation or block, and arrhythmias associated with coronary ligation were increased (7). In cats

T-wave alternans was induced by asymmetric stimulation of one or both sympathetic ganglia for 5 to 30 seconds, the effect persisting for 20 to 30 seconds after cessation of stimulation (8). The similarity between these experimental results and the clinical profile of LQTS led to the suggestion that the fundamental abnormality in the syndrome could be an asymmetrical defect of cardiac sympathetic control (9). "Taken together the available clinical and experimental data seem to strongly support the hypothesis of a congenital imbalance in the cardiac sympathetic innervation, with a dominance of left-sided nerves that is likely secondary to a lower than normal right cardiac sympathetic activity" (10).

The arrhythmias in LQTS could often be prevented by treatment with beta-blockers, which raised this question: If the problem was a deficit of sympathetic control, how could it be solved by reducing sympathetic activity further still? An explanation was offered by reference to the effects of prolonged β-adrenoceptor blockade (2).

Long-term treatment of rabbits with beta-blockers induces several adaptive changes in the myocardium, which persist long after the drugs have been eliminated from the body. This "adaptation syndrome" included a prolongation of action potential duration and QT, and both these effects have been shown to occur in man. These observations could explain both the prevalence of arrhythmias in the long QT syndrome and the success of anti-sympathetic therapy by left stellate ganglionectomy or beta-blockade. The long QT syndrome is said to be associated with a preponderance of sympathetic innervation from the left stellate ganglion, and a deficit on the right, accounting for a reduced tachycardia on exercise in patients, since the normal sinoatrial node is primarily innervated from the right side. The ventricular fibers deprived of their (right) sympathetic innervation would adapt by having very long action potential durations (hence the long QT). During exercise or emotion the myocardial regions supplied by the left stellate would receive excessive sympathetic drive, since they would have to compensate for the failure of the non-innervated areas to increase their activity. The myocardium would thus be subjected to two highly arrhythmogenic influences, the juxta-position of shortened action potential duration in the innervated fibers next to adapted and unresponding long action potential durations in the remainder, and the probability of the awakening of subsidiary pacemaking cells by the aggravated adrenergic stimulation. Aboli-tion of the excessive sympathetic drive by surgery or by adrenergic blockade would reduce the heterogeneity of action potential duration and though peak performance might be diminished, this would be a reasonable price to pay for a lessened risk of sudden death (2).

The mortality in untreated patients with LQTS is extremely high—71% (10). Although beta-blockade was able to control symptoms in 76% to 80%, the remaining 20% to 24% were still at great risk, which could be reduced by surgical removal of the first four or five left sympathetic ganglia. If the upper part of the left stellate was not resected, Horner's syn-drome could usually be avoided. Schwartz et al. (15) have kept an international register of LQTS patients, and reported on 85 subjects who had undergone left cardiac sympathetic denervation (LCSD). The time interval between the first cardiac event and the opera-tion was 5.6 ± 6.1 years, and the follow-up period after operation was 5.9 ± 5.7 years. The mean age at surgery was 20 ± 13 years. The 5-year survival was 94% worldwide, but for the 25 patients treated in Milan it was 100% (11). "The most likely mechanisms for the pro-tective effect of LCSD are the electrophysio-logic consequences of a markedly reduced release of norepinephrine in localized areas of the ventricles" (11). The question remains why LCSD should succeed when beta-block-ade has failed. "The efficacy of sympathetic denervation in patients who continue to have syncope or cardiac arrest despite full-dose beta-blockade points strongly to the impor-tance of an α-adrenergic mechanism in the arrhythmias of LQTS" (11). Selective α-adrenoceptor stimulation was shown to cause a prolongation of cardiac APD (see Chapter 3, Ref. 14). Combined α- and β-adrenoceptor blockade has not been employed for the treat-ment of LQTS patients. Labetolol, which has both β- and α-adrenergic blocking effects,

was not used because there is little experience with its use in patients with the long QT syn-drome (12).

Although a predominant characteristic of LQTS is, of course, prolongation of QT, this was not always much beyond normal limits. The T-wave could be bifid or notched. Heart rate was slower than normal at rest in chil-dren, and there could be occasional long pauses and syncope. In addition various degrees of AV block were not uncommon. To avoid these problems Eldar et al. (12) rou-tinely implanted pacemakers in all LQTS patients, setting the basic rate high enough to reduce QT_C to <440 ms, which necessitated heart rates of 70 to 125 beats·min-1 (mean 83 $\pm$ 12). Patients were also treated with beta-blockers. In a report of a follow-up of at least 6 months on 21 patients "who had maintained both beta-blocker therapy and consistent long-term pacing, only two had a cardiac event" (12). In one a dual-chamber pace-maker was inserted to exclude intermittent atrioventricular (AV) block. The other had a cervicothoracic sympathectomy and a dual-chamber pacemaker, and subsequently remained event-free for 6 months (12).

The origin of the long QT may be differ-ent in LQTS from that observed after myocardial infarction (MI), but the cause of the arrhythmias is the same—dispersion of ventricular repolarization times. In a com-parison between 15 healthy volunteers as controls and 29 patients with Romano-Ward type LQTS, QT dispersion was measured from digitized 12-lead ECGs as (a) the dif-ference between the longest and shortest QT intervals in each lead, and (b) the relative dispersion (i.e., the standard deviation of QT divided by the average QT multiplied by 100). Both indices of dispersion were higher in the LQTS patients than in control sub-jects; also, patients not responding to beta-blockers had a significantly higher disper-sion of repolarization than responders. The LQTS patients who did not respond to beta-blockade underwent left cardiac sympathetic denervation and thereafter remained asymp-tomatic (mean follow-up, 5 ± 4 years). In this group, dispersion of repolarization was significantly reduced by the surgical denervation to values similar to that of the responders to beta-blockade (13).

The effect of beta-blockers in reducing QT_C dispersion in patients with LQTS during exercise was confirmed in seven subjects by Statters et al. (14).

THE GENETIC DEFECT IN LQTS

In a prospective study of the families of patients with LQTS, the first member to be identified (the proband) was usually brought to medical attention because of a syncopal episode in childhood.

"Probands (n =328) were younger at first contact (21 ± 15 years), more likely to be female (69%), and had a higher frequency of preenrollment syncope or cardiac arrest with resuscitation (80%), congenital deafness (7%), a resting heart rate less than 60 beats/min (31%), $QT_C \geq 0.50$ $sec^{0.5}$ (52%), and a history of tachyarrhythmia (47%), than other affected ($n = 688$) and unaffected ($n = 1004$) family members. Arrhythmogenic syncope often occurred in association with acute physical, emotional, or auditory arousal. The syncopal episodes were frequently misinterpreted as a seizure disorder (15).

85% of probands had family members with $QT_C > 0.44$, and only 10% had no family history of syncope or sudden death before the age of 50; 47% had one or more syncopal episodes associated with stress or emotion (15)."

The above study provides overwhelming support to the early conclusion that the LQTS is an inherited disorder of repolarization, causing arrhythmias associated with sympathetic stimulation. Repolarization can only occur if K^+ conductance predominates, because E_{Na} and E_{Ca}, though changing during the action potential, are always positive, and E_{Cl}, though more stable, is usually not more negative than -40 mV. Thus, for the membrane to repolarize to fully negative diastolic potentials, potassium channels must open sufficiently for K^+ current to prevail.

MOLECULAR STRUCTURE OF POTASSIUM CHANNELS

Since the cloning of the first inwardly rectifying K^+ channel in 1993, a family of related clones has been isolated, with many members expressed in the heart (see Chapter 4, Ref. 33), in which there are at least six types of K^+ channel genes (16). When the amino-acid sequence is established the tertiary structure has to be deduced from the positions of predominantly hydrophilic and hydrophobic series.

The membrane-embedded sequence is mapped as 6 α-helices (S1–S6), each ~20 residues in length, which zig-zag through the membrane. S4 is highly conserved and has a net positive charge arising from an Arg/Lys-X-X-Arg/Lys repeat, where the X's are usually hydrophobic. S4 is thought to be important for voltage sensing. A seventh transmembrane stretch called H5 lies between S5 and S6, and is thought to form the ion conduction pathway or pore, and may be referred to as the P region. It may be assumed that four K^+ channel subunits assemble to form the complete voltage-dependent K^+ channels (16).

Two mechanisms for inactivation have been described, the "ball-and-chain" and core (C-type).

An NH_2-terminus stretch of 19 residues was shown to produce fast inactivation (N-type). In the context of the ball-and-chain metaphor used for [intracellular] tryptic removal of inactivation in Na^+ channels, a similar metaphor was adopted for K^+ channels.... Internal but not external tetraethylammonium (TEA) blocked N-type inactivation, an effect that was explained as a simple competition between TEA and the inactivating ball for a shared internal binding site. In contrast, external but not internal TEA blocked C-type inactivation. Competition for a common binding site was again suggested, but in this case a "foot-in-the-door" model was used (16).

Detailed molecular modeling of different types of K^+ channels still leaves open the question of how structure relates to function. What advantage is conferred by possession of both fast and slow components of delayed rectification? Why do Purkinje cells and epicardial cells, but not endocardial cells, have I_{TO} channels? Until such questions are answered it is impossible to predict what overall effect, antiarrhythmic or proarrhythmic, would be exerted by selective blockade of a particular channel.

ABNORMALITIES OF K+ CHANNELS IN LQTS

In 1991 a tight linkage was discovered between autosomal dominant LQT and chromosome 11p15.5, and because the first seven families studied were similarly linked, autosomal dominant LQT was thought to be genetically homogeneous (LQT1). In 1993, however, families with LQTS were identified without linkage to 11p15.5, and two additional loci were discovered, LQT2 on chromosome 7q35-36 (nine families), and LQT3 on chromosome 3p21-24 (three families). Further studies led to identification of the human ether-a-go-go gene (HERG) with LQT2 (17). As for LQT3, there is evidence that the patients with this mutation have an abnormality of a sodium channel, SCN5A (18). Yet a fourth mutation LQT4 is located on chromosome 4q25-26. Meanwhile positional cloning methods established the chromosome 11 linked gene LQT1 as KVLQT1, which is strongly expressed in the heart and encodes a protein with the structural features of a voltage-gated potassium channel. KVLQT1 mutations were present in affected members of 16 families with LQTS arrhythmias (19).

The abnormality associated with the LQT3 mutation causes failure of the SCN5A sodium channel to inactivate normally, leading to multiple openings that prolong the plateau sodium current. In contrast, the LQT2 (HERG) associated defect reduces current through delayed rectifier channels, postponing the onset of repolarization (20). The HERG gene codes the fast component of delayed rectification, and when the channel is expressed in *Xenopus* oocytes it is blocked by methanesulfonamide class 3 agents (21). In similar preparations it was demonstrated that the inactivation mechanism in the HERG channel is not of a ball-and-chain type, but a C-type (22).

In a group of LQTS patients with a specific HERG mutation (Ala561Thr), "a distinctive biphasic T-wave pattern was found in the left precordial leads of all affected subjects, regardless of age, gender, and beta-blocking therapy" (23). In seven adult patients with LQTS identified as associated with a chromosome 7–linked mutation, raising serum K+ concentration by spironolactone, coupled with potassium administered IV and orally, corrected "abnormalities of repolarization duration, T-wave morphology, QT/RR slope, and QT dispersion. The short-term nature of this study does not permit conclusions to be made about the possible benefits of long-term potassium therapy" (24).

The incidence of arrhythmias associated with LQTS is greater in females than males. In a study of 14,379 subjects from birth to age 75, the QT interval, measured as an index correcting for heart rate by an algorithm more sophisticated than the Bazett equation, was found to be about 20 ms longer in males than in females after puberty (25). Direct experimental evidence for the effect of sex hormones in prolonging QT and downregulating K+ channels was obtained in rabbits (26).

IMPLICATIONS FOR THERAPY

Although patients with LQTS usually have lower than normal heart rates and a reduced response to exercise, a common feature is that arrhythmias are precipitated by stress, physical or emotional. It was concluded that the genetic defect probably resided in the sympathetic nervous system, causing a preponderance of left- over right-sided activity. Now that the genetic abnormalities have been located in genes coding for ion channels, it may be presumed that affected channels induce an abnormal response to adrenergic stimulation. β-Adrenoceptor–mediated activity both increases heart rate (which shortens APD) and shortens APD directly, by augmenting potassium conductance. The cause of arrhythmias in LQTS may be attributed to a failure of cells with deficient delayed rectifier channels to keep pace when repolarization is accelerated under sympathetic drive, leading to increased dispersion of repolarization. The delay in repolarization from the plateau, due to persisting inward sodium current or defective outward rectifier current, makes repolarization more dependent on the inward rectifier, I_{K1}. LQTS patients are, therefore, more vulnerable to a fall in serum K^+, because E_K becomes more negative, and I_{K1} only switches on as E approaches E_K. Strong adrenergic activation lowers serum K^+, because β_2-adrenoceptors activate the Na/K pump throughout the musculature. Optimum therapy, therefore, may still be restriction of sympathetic activity, by beta-blockade or surgery. In families with the LQT3 mutation, in whom repolarization is delayed by excessive plateau sodium current, a class 1 agent could be of additional benefit.

TORSADE DE POINTES

In 1966, in an article more often cited than studied, Dessertenne (27) presented the case history of an 80-year-old woman with atrial fibrillation and ventricular tachycardia of an unusual type. He was concerned to distinguish between fusiform complexes associated with ventricular fibrillation (VF) and a different kind in which the ventricular deflections were pointed, with the points directed upward in one group and downward in the next. He gave this sequence of alternating tachycardia the name "torsade de pointes." Describing the characteristic tracing (a detail of which is reproduced in Fig. 1) he wrote, "The 3 first torsades are particularly interesting. Each lasts about 3 seconds. Examining them closely one sees that in lead III the components of the first have the point directed upward, those of the second have the point downward, and those of

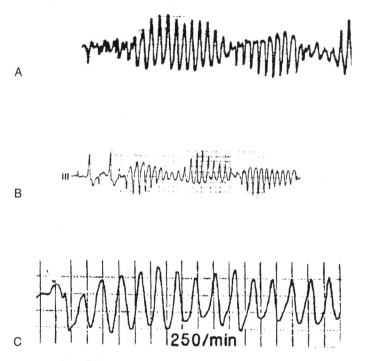

FIG. 1. A: ECG illustrating pointed deflections at first upward, then downward, then upward again *(tantôt en haut, tantôt en bas)*. **B:** ECG with characteristics similar to those described by Dessertenne. **C:** ECG described as *torsade de pointes,* but in which the excursions are rounded, not pointed, and in which there is no evidence of *deux foyers.*

the third have the point once more directed upward." He recalled that fusiform complexes were customarily explained as produced by "combinations of electrical activity from several sources, sometimes in phase, sometimes of opposite phase." When the excursions alternated direction above and below the "reference line," he concluded that they must be originating in turn from two separate sources of opposite orientation, "*deux foyers opposés.*" The individual excursions were truly pointed, not rounded, and all the points were in the same direction; he distinguished clearly between "*une pointe et un arrondi, toutes les pointes étant dans le même sens*" (27).

A decade later, in a scholarly discussion of torsades de pointes, Krikler and Curry (28) noted, "The specific findings consist of paroxysms of ventricular tachycardia in which the QRS axis undulates over runs of 5 to 20 beats, with definite changes in direction." These features are well illustrated in an ECG from which a detail is reproduced in Fig. 1B.

In the etiology of torsade de pointes,

of major importance are bradycardias, very often high-grade atrioventricular block or sino-atrial block. It is absolutely vital not to interpret the appearance as representing ventricular hyperexcitability, for quinidine and similar agents given on this assumption may disastrously aggravate the arrhythmia. The underlying situation must be corrected—for example potassium should be replaced, and quinidine or other drugs withdrawn. When heart-block is the underlying cause cardiac pacing is the definitive therapy. Torsade de pointes requires urgent treatment quite different from that of classical ventricular tachycardia (28).

In 1989, in an assessment of whether "the electrophysiologic substrate of this arrhythmia [torsade de pointes] is created by an increased dispersion of repolarization," Surawicz (29), confirming Dessertenne's definition, stated that "to satisfy the description of torsade de pointes, the axis of the QRS complex must change direction after a certain number of complexes, as if the complex rotated around the baseline. The electrical activity in this condition is less disorganized than in ventricular

fibrillation." The description of torsade de pointes by Dessertenne himself and the above authors is so clear that it is regrettable that the term has often been applied to quite different conditions.

MISNOMERS

Drug-associated ventricular arrhythmias have frequently been referred to as torsade de pointes even when they do not exhibit the appropriate characteristics. For example, an imipramine-induced arrhythmia, a detail of which is depicted in Fig. 1C, was described as "torsades de pointes," although there are no points, excursions in both directions being rounded (arrondis), and there is no rotation of the axis about the baseline (30). The authors cited the paper by Dessertenne but referred to "torsades de pointes" as "showing the characteristic undulating pattern or 'twisting of the points.' The points refer to the tips of the QRS complexes." This description does not conform to that of Dessertenne, but has evidently been adopted by many subsequent authors, which is unfortunate because to give the same name to quite different entities causes confusion. It has been suggested that the term *proarrhythmia* be reserved for arrhythmias associated with therapeutic concentrations of drugs, and that arrhythmogenesis be used when higher concentrations are involved (31).

Another misnomer is "acquired LQTS." Since LQTS is an inherited condition it cannot be acquired. Class 3 agents prolong QT interval and are sometimes proarrhythmic, but to refer to such a combination as an acquired long QT syndrome can only create confusion. The suggestion that patients in whom arrhythmias occur in association with antiarrhythmic drug therapy may be especially sensitive because of an inherited abnormality similar to those observed in LQTS has been shown to be without foundation (32).

REFERENCES

1. Schwartz PJ, Wolf S. QT interval prolongation as a predictor of sudden death in patients with myocardial infarction. *Circulation* 1978; 57:1074–1077.

2. Vaughan Williams EM. QT and action potential duration. *Br Heart J* 1982; 47:513–514.
3. Jervell A, Lange-Nielson F. Congenital deaf-mutism, functional heart disease with prolongation of the QT interval, and sudden death. *Am Heart J* 1957; 54:59–68.
4. Romano C, Gemme G, Pongiglione R. Aritmie cardiache rare dell eta pediatrica. *Clin Pediatr* 1963; 45:65.
5. Ward OC. A new familial cardiac syndrome in children. *J Ir Med Assoc* 1964; 54:103–106.
6. Ward OC. The inheritance of paroxysmal tachycardia. In: Sand E, Flensted-Jensen E, Olsen KH, eds. *Symposium on cardiac arrhythmias.* Södertälj, Sweden: AB Astra, 1970; 92–107.
7. Schwartz PJ, Stone HL, Brown AM. Effects of unilateral stellate ganglion blockade on the arrhythmias associated with coronary occlusion. *Am Heart J* 1976; 92:589.
8. Schwartz PJ, Malliani A. Electrical alternation of the T-wave: clinical and experimental evidence of its relationship with the sympathetic nervous system and with the long QT syndrome. *Am Heart J* 1975; 89:45–50.
9. Schwartz PJ, Periti M, Malliani A. The long QT syndrome. *Am Heart J* 1975; 89:378–390.
10. Schwartz PJ. Idiopathic long QT syndrome: progress and questions. *Am Heart J* 1984; 108:399–411.
11. Schwartz PJ, Locati EH, Moss AJ, Crampton RS, Trazzi R, Ruberti U. Left cardiac sympathetic denervation in the therapy of congenital long QT syndrome. *Circulation* 1991; 84:503–511.
12. Eldar M, Griffin JC, van Hare GF, Witherell C, Bhandari A, Benditt A, Scheinman MV. Combined use of beta-adrenergic blocking agents and long-term cardiac pacing for patients with the long QT syndrome. *J Am Coll Cardiol* 1992; 20:830–837.
13. Priori SG, Napolitano C, Diehl L, Schwartz PJ. Dispersion of the QT interval. A marker of therapeutic efficacy in the idiopathic long QT syndrome. *Circulation* 1994; 89:1681–1689.
14. Statters DJ, Panagos A, Fei L, Rowland E, Malim M, Camm AJ. How does beta-blockade protect patients with congenital long QT syndrome ? *Br Heart J* 1995; 73(suppl 3):58.
15. Moss AJ, Schwartz PJ, Crampton RS, Tzivoni D, Locati EH, MacCluer J, Hall J, Weitkamp L, Vincent M, Garson A, Robinson JL, Benhorin J, Choi S. The long QT syndrome. Prospective longitudinal study of 328 families. *Circulation* 1991; 84:1136–1144.
16. Taglialatela M, Brown AM. Structure correlates of K^+ channel function. *News Physiol Sci* 1994; 9:169–173.
17. Curran ME, Splawski I, Timothy KW, Vincent GM, Green ED, Keating MT. A molecular basis for cardiac arrhythmia: HERG mutations cause long QT syndrome. *Cell* 1995; 80:795–803.
18. Wang Q, Shen J, Splawski I, Atkinson D, Li Z, Robinson JL, Moss AJ, Towbin JA, Keating MT. SCN5A mutations associated with an inherited cardiac arrhythmia, long QT syndrome. *Cell* 1995; 80:805–811.
19. Wang Q, Curran ME, Splawski I, Burn TC, Millholland JM, van Raay TJ, Shen J, Timothy KW, Vincent GM, de Jager T, Schwartz PJ, Towbin JA, Moss AJ, Atkinson DL, Landes GM, Connors TD, Keating MT. Positional cloning of a novel potassium channel gene: KVLQT1 mutations cause cardiac arrhythmias. *Nature Genet* 1996; 12:17–23.
20. Keating MT, Sanguinetti MC. Molecular genetic insights into cardiovascular disease. *Science* 1996; 272:681–685.
21. Spector PS, Curran ME, Keating MT, Sanguinetti MC. Class III antiarrhythmic drugs block HERG, a human cardiac delayed rectifier K^+ channel. *Circ Res* 1996; 78: 499–503.
22. Schönherr R, Heinemann SH. Molecular determinants of activation and inactivation of HERG, a human inward rectifier potassium channel. *J Physiol* 1996; 493:635–642.
23. Dausse E, Berthet M, Denjoy I, André-Fouet X, Cruaud C, Bennaceur M, Fauré S, Coumel P, Schwartz K, Guicheney P. A mutation in HERG associated with notched T-waves in long QT syndrome. *J Mol Cell Cardiol* 1996; 28:1609–1615.
24. Compton SJ, Lux RL, Ramsey MR, Strelich KR, Sanguinetti MC, Green LS, Keating MT, Mason JW. Genetically defined therapy of inherited long QT syndrome. *Circulation* 1996; 94:1018–1022.
25. Rautaharju PM, Zhou SH, Wong S, Calhoun HP, Berenson GS, Prineas R, Davignon A. Sex differences in the evolution of the electrocardiographic QT interval with age. *Can J Cardiol* 1992; 8:690–691.
26. Drici MD, Burklow TR, Haridasse V, Glazer RI, Woosley RL. Sex hormones prolong the QT interval and down regulate potassium channel expression in the rabbit heart. *Circulation* 1996; 94:1471–1474.
27. Dessertenne F. La tachycardie ventriculaire à deux foyers opposés variables. *Arch Mal Coeur* 1966; 59:263–272.
28. Krikler DM, Curry PVL. *Torsade de pointes,* an atypical ventricular tachycardia. *Br Heart J* 1976; 38:117–120.
29. Surawicz B. Electrophysiologic substrate of torsade de pointes: dispersion of repolarization or early after depolarization? *J Am Coll Cardiol* 1989; 14:172–184.
30. Jackman WM, Friday KJ, Anderson JL, Aliot EM, Clark M, Lazzara R. The long QT syndrome: a critical review, new clinical observations and a unifying hypothesis. *Prog Cardiovasc Dis* 1988; 31:115–172.
31. Somberg JC. Proarrhythmia or arrhythmogenicity. *Am J Ther* 1995; 2:149.
32. Wei J, Warhen M, Murray K, Daw R, Roden D, George L. Absence of HERG and SCN5A mutations in acquired long QT syndrome. *Circulation* 1996; 68th Sci Session Abstr. No. 1305.

6

Class 4 Antiarrhythmic Action

E. M. Vaughan Williams

SKELETAL AND CARDIAC MUSCLE

The diameter of a single cell is limited by its metabolic demands; hence, the necessity for multicellular organisms. Rapid locomotion requires large structures capable of rapidly generating and transmitting great force. Skeletal muscle fibers are up to 100 μm in diameter, and have very low average oxygen uptake because they are inactive most of the time, especially in industrial human populations, standing or seated nearly all day. Mitochondria in skeletal muscle are sparsely distributed and perform a "battery-charger" function, replenishing high-energy phosphate and glycogen consumed by infrequent bursts of activity. A sprinter does not use the oxygen breathed during a race for body muscle function, being able to maintain a speed of up to 10 m·s⁻¹ for several seconds anaerobically. The average speed for the 200 m record is higher than for 100 m, because the athlete has a flying start for the second 100 m. Conversely, a marathon runner has to pace himself carefully to spend only 1/26th of his energy reserves per mile.

More than 60 years ago A. V. Hill emphasized that the coupling between the surface action potential of a skeletal muscle fiber and its contractile elements could not be due to diffusion of a messenger such as calcium, because it would be far too slow, and it was not until the advent of electron microscopy that the T-tubular system was discovered, providing each sarcomere with its own individual electrical field signal as the surface action potential spreads into the interior. It is still uncertain how this signal is transferred from the tubule to release calcium from the sarcoplasmic reticulum (SR), and here, too, an electrical field may be involved. If there is a potential difference between the inside of the SR and the cell cytosol, electrotonic discharge of the SR membrane capacity could involve the net transfer of only a small number of ions. It has been suggested that actin itself may exhibit cable-like conduction properties (1).

Extracellular calcium concentration has little influence on the force of skeletal muscle contractions, calcium being recycled intracellularly. The rate of removal of SR-released calcium varies greatly, being slowest in postural muscles that provide a steady tetanic contraction at a fusion frequency of 30 action potentials per second, and fastest in extraocular muscles, the fusion frequency of which exceeds 300 s⁻¹.

In contrast, cardiac muscle runs a marathon for life. Mitochondria occupy 40% of the intracellular volume, and cell diameter is 7 to 14 μm. Even in extreme hypertrophy a diameter of 20 μm is exceptional. Energy reserves of glycogen are limited, but adenosine triphosphate (ATP) concentration is about 7 mM. The immediate response to hypoxia is depression of contraction, so that depletion of energy stores is delayed. Contractile force is highly sensitive to extracellular calcium.

VOLTAGE-GATED CALCIUM CHANNELS

The existence of a second inward current depolarizing cardiac muscle and carried by ions other than sodium, probably calcium, was recognized in 1958 (see Chapter 1, Ref. 3), but the charge carrier was not properly identified by voltage clamp experiments for a decade. The second, or slow, inward current, I_{si}, was later shown to have two components, one activated by depolarization to about -50 mV and rapidly inactivated, and therefore transient, I_{Ca-T}, and the other activated at voltages positive to -40 mV, and inactivated more slowly, and therefore longer-lasting, I_{Ca-L}. The latter was also inactivated by rising intracellular calcium concentration (negative feedback). Several types of Ca channels have been cloned and expressed in *Xenopus* oocytes. The molecular structure of L-type cardiac calcium channels broadly resembles that of Na and K channels in possessing four groups of six main transmembrane columns, S1 to S6, with a pore loop between S5 and S6. The ion channel pore is formed by four structures from which glutamate residues point inward to provide Ca^{2+} selectivity by binding two ions in series (2).

The whole amount of any ion that enters the cell during systole must be removed well before the end of diastole, to avoid indefinite accumulation, although the concentration at which a steady state is achieved may vary from time to time. Intracellular sodium and potassium concentrations are controlled by the Na/K pump. Intracellular calcium has a number of destinations, most of them temporary—mitochondria, in which they have a controlling function; troponin; several calcium-trapping proteins; and the SR, into which Ca is actively pumped for subsequent release. All these structures are intracellular, and cannot shift calcium ions to the exterior, a process that must be achieved rapidly because $[Ca^{2+}]_i$ must fall below 10^{-7} M to permit muscular relaxation.

SODIUM/CALCIUM EXCHANGE

The energy for calcium extrusion is provided indirectly by the Na/K pump, which establishes a low intracellular Na^+ concentration (about 10 mM). A sodium/calcium exchange mechanism permits three Na^+ ions to diffuse into the cell in exchange for one outward Ca^{2+} ion, providing a depolarizing current that helps to maintain the action potential plateau. Calcium entering through voltage-gated channels stimulates further calcium release from the SR by a mechanism that can be completely blocked by ryanodine 50 mM, and that can be simulated by caffeine. SR contains K^+ channels that can open or close in response to changes in the relative Ca^{2+} concentrations inside and outside the SR membrane (3). In feline atrial myocytes "voltage-dependent Ca^{2+} entry triggers Ca^{2+} release from peripheral coupling SR that subsequently induces further Ca^{2+} release from stores in more central regions of the myocyte. The finding that cat atrial myocytes lack T-tubules demonstrates the functional importance of Ca^{2+} release from extended junctional (corbolar) SR in these cells. In ventricular myocytes the well-developed T-tubular system ensures a spatially homogeneous Ca^{2+} entry through voltage-gated Ca^{2+} channels into the entire cell volume" (4).

In rat ventricular myocytes there is evidence of a nonuniform distribution of Ca^{2+}_i. In low extracellular calcium solutions (1 mM) waves of spontaneous Ca^{2+} release from the SR occurred in 70% of cells, and the concentration of the subsarcolemmal calcium changed more rapidly than in the cell interior, so that a wave of calcium traveled longitudinally, as if a corridor existed between the membrane and a barrier separating it from the deeper cytosol (5). In a series of voltage clamp experiments in guinea pig ventricular myocytes Sipido et al. (6) concluded that although $[Na^+]_i$ did not cause release of calcium from the SR, it could influence uptake of calcium by the SR via the Na/Ca exchange mechanism, and that the critical feature controlling the exchange was the Na^+ concentration in a region 10 nm deep immediately under the sarcoplasmic membrane.

When a cell is depolarized the force driving Na^+ inward is reduced, and that driving Ca^{2+} outward is increased, so that the Na/Ca exchange mechanism can go into reverse, 3 Na^+ ions being extruded in exchange for 1 Ca^{2+} entering. This provides a repolarizing current, but increases the subsarcolemmal calcium, leading to speculation that reverse Na/Ca exchange may contribute to the Ca^{2+} activating a normal contraction. Evans and Cannell (7) concluded that calcium entry through L-type channels was the major trigger for Ca^{2+} release from SR, because when the channels were blocked by 10 mM verapamil, activation of I_{Na} by clamping to -40 mV did not induce a calcium transient. Nevertheless, even if Na^+ does not release calcium directly from the SR, it could influence the amount released, by controlling the rate of uptake into the SR relative to the rate of extrusion of Ca^{2+} by the Na/Ca exchanger. Stimulating rat ventricular myocytes with a series of short depolarizing pulses caused less and less uptake of Ca^{2+} by the SR, because less Ca^{2+} had time to enter via L-type channels. Conversely, long pulses caused more and more uptake, calcium becoming available not only from L-type channels but also from Na/Ca exchange (8).

Wasserstrom and Vites (9) concluded that in rat ventricular myocytes calcium entry through L-type channels (which could be blocked by nifedipine) provided only about half of the contraction-coupling calcium, proposing that the remainder was supplied by Na/Ca exchange. In rabbit heart cells reduction of I_{Ca-L} to 1/10th of control values (by nifedipine) reduced the Ca^{2+} transient, measured by a light-emitting probe (fura-2), by only about half. The nifedipine-independent component was abolished by 5 mM of nickel, which blocks Na/Ca exchange. The conclusion was that Na/Ca exchange does partially control the amount of Ca^{2+} taken up by the SR (10). A similar conclusion was drawn from experiments on guinea pig cardiac myocytes. After block of I_{Ca-L} by nifedipine, 24% of the total calcium transient, measured by an Indo-1 probe, could still be elicited by an 800-ms pulse with $[Na^+]_i$ fixed at 10 mM. The higher the $[Na^+]_i$ and the longer the pulse, the larger the L-type–independent calcium transient. "The results support the idea that both I_{Ca-L} and a second mechanism are able to trigger SR release and the resulting Ca^{2+}_i transient. The second SR trigger mechanism is Ca^{2+} entry via reverse Na^+-Ca^{2+} exchange, elicited by a step change in membrane potential" (11).

Since the early work of Bowdich a connection between stimulation frequency and force of contraction in cardiac muscle has been recognized and various models of the strength-interval relation have been proposed. In rabbit atrial muscle the relation is bimodal (12). In guinea pig ventricular myocytes an increase in frequency caused $[Na^+]_i$ to rise, which in turn increased reverse Na/Ca exchange, leading to greater Ca^{2+} uptake and release by the SR, augmenting contractions (13). This explains the "positive staircase," the larger contraction associated with higher frequency. It does not explain, however, why after very long pauses contractions are increased in atria (12), though diminished in ventricles. It may be that in ventricular muscle, at rest, calcium is slowly lost from the SR to the T-tubules, whereas in resting atria, which lack T-tubules, there may be a net accumulation, perhaps from the extracellular space.

In rat sarcolemmal vesicles Na/Ca exchange can be increased by α-adrenoceptor stimulation, by angiotensin II, and by endothelin I, but not by β-adrenoceptor agonists or by carbachol (14). The inward exchange current in guinea pig ventricular myocytes is selectively blocked by a new compound No. 7943 (IC_{50} = 0.32 µM). Much higher concentrations reduced outward exchange current, inward sodium current, L-type calcium current, and I_{K1} (IC_{50} = 17, 14, 8, and 7 µM, respectively) (15). The relevance of such experiments to cardiac function is unclear because the effect of No. 7943 on the action potential and contractions of isolated beating cardiac preparations was not recorded, and in spite of the evidence that reverse Na/Ca exchange can contribute to Ca^{2+} release from the SR in specialized conditions in ventricular myocytes of various species, it is unknown whether such a current controls contraction in human ventricles.

CALCIUM ANTAGONISTS

The concept of calcium antagonism was originally concerned with inhibition of the Ca-activated adenosine triphosphatase (ATPase). In their review in 1966 Fleckenstein et al. (16) stated, "As to the inhibitors of utilization, they seem either to block the movement of Ca from the excited membrane to the myofibrils or to compete with Ca for the active sites in the contractile system where ATP is split. The list of substances cited as capable of producing contractile failure of the heart by acting as Ca-antagonists in excitation contraction coupling included a number of beta-receptor blocking substances and related compounds (isoptin, segontin). Some local anesthetics have similar effects" (16).

There was no mention of a second inward current, or of its blockade by calcium antagonists.

Verapamil, like amiodarone, was originally introduced as a coronary dilator for the treatment of angina pectoris. It did not block the effects of isoprenaline on ventricular stroke volume or on bronchial muscle, and it was not a competitive antagonist of catecholamines (17). Soon after its introduction verapamil was discovered to have antiarrhythmic properties (18). The idea of a fourth class of antiarrhythmic action due to the block of the second inward current carried by calcium ions was based on electrophysiologic studies in spontaneously beating rabbit atria and in paced papillary muscles (see Chapter 1, Ref. 14).

Verapamil and Diltiazem

Calcium ions are ubiquitous intracellular messengers for so many functions that attempts to control an individual calcium-activated mechanism depend on finding a highly selective inhibitor or agonist of the particular function. Verapamil and diltiazem may be considered to be antiarrhythmic drugs because, although they dilate coronary arteries and other blood vessels, and are, therefore, of value in the treatment of angina pectoris and hypertension,

they do possess some relative selectivity for calcium channels in the sinoatrial node (SAN) and atrioventricular node (AVN).

In the AVN, as in the SAN, there are two main types of cell. The ovoid node (N) cells with low resting potentials contain few sodium channels, and resemble the p cells of the SAN, which lack I_{K1} channels. In both N and p cells, calcium ions are the depolarizing charge carriers. The other type of AVN cell (N-H), rod-shaped and containing more sodium channels, corresponds to the transitional cell of the SAN. N-H cells connect the N cells to the Purkinje cells of the His bundle, and have low resting potentials because they also lack I_{K1} channels (19). Sodium channels, absent from N cells, become more numerous in the progression through N-H to Purkinje cells, but dependence on calcium as charge carriers renders the AVN especially susceptible to conduction delay by drugs blocking calcium channels. Verapamil and diltiazem are clinically useful for the control of supraventricular arrhythmias, especially those involving reentry in and around the nodes. Selectivity for cardiac calcium channels, however, makes these drugs less safe in the presence of β-blockade.

Dihydropyridines

The introduction of nifedipine heralded a new generation of calcium antagonists, with much greater selectivity for calcium channels in blood vessels than in the heart. The hypotensive effects of these drugs provoked reflex increases in sympathetic activity, so that, in spite of a direct bradycardic action on the SAN, heart rate did not fall, or was increased. Dihydropyridines can therefore be safely employed in combination with beta-blockers, but can no longer be regarded as antiarrhythmic drugs. There are now numerous dihydropyridine calcium antagonists, varying in subsidiary features such as absorption, distribution, and duration of action, which have frequently been reviewed as new compounds have been introduced (20), and have recently been classified by structure (Fig. 1) and binding properties (21).

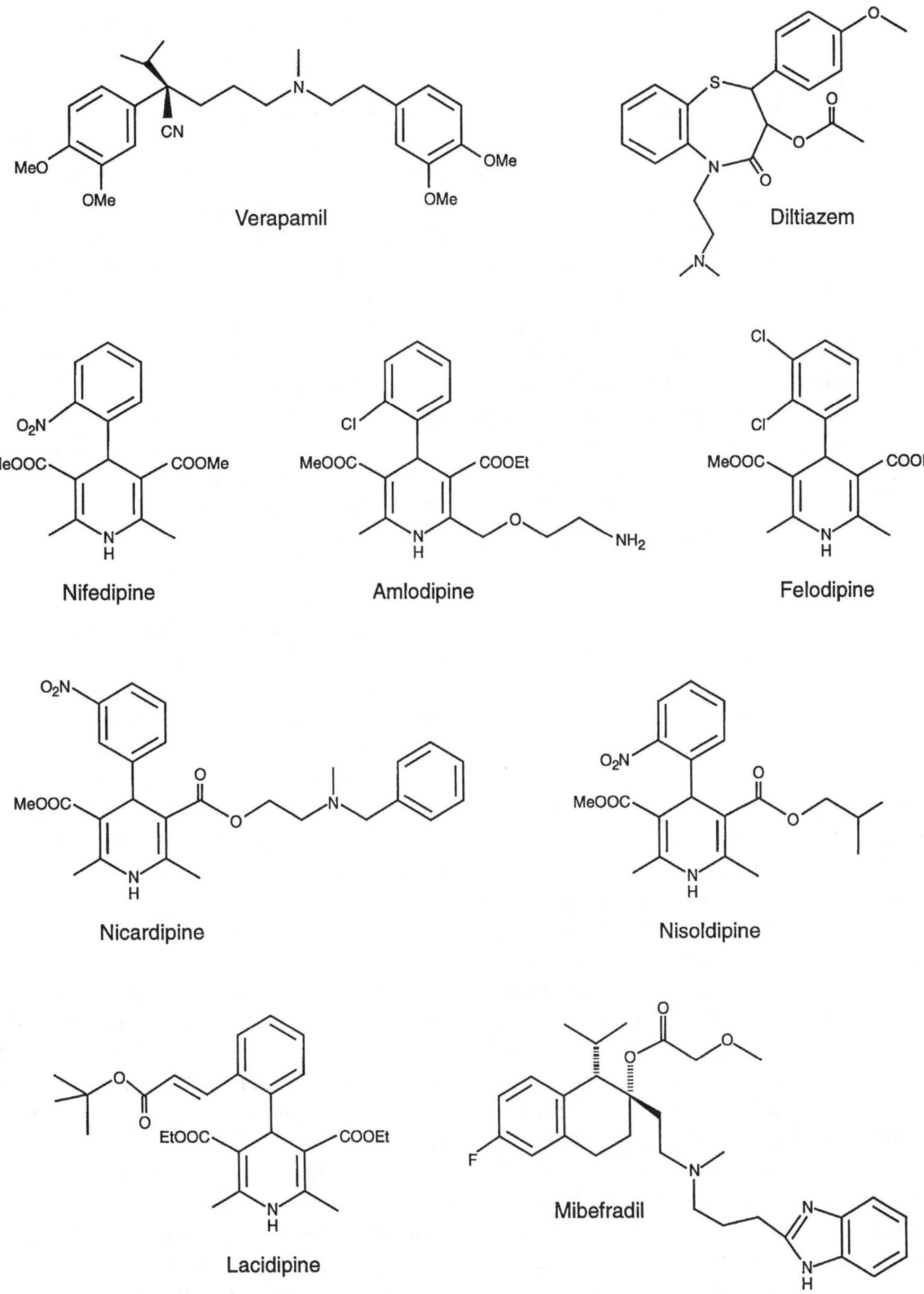

FIG. 1. Chemical structures of some calcium antagonists.

Mibefradil

Mibefradil is a new calcium antagonist that is not a dihydropyridine, but appears to have generally similar properties. It blocks I_{Ca-T} as well as I_{Ca-L}, and is sufficiently selective for vascular calcium channels to have little negative inotropic action at therapeutic concentrations. Consequently mibefradil could be used to reduce vascular resistance without endangering the myocardium in heart failure. Ninety percent is absorbed on oral administration and it has an elimination half-life of 17 to 25 hours. Mibefradil has no class 1 activity (IC_{50} for block of I_{Ca-L} is 0.2 μM, and for block of I_{Na} is 55 μM). It does not prolong PR or QT intervals (22). It is improbable that mebefradil would exhibit any antiarrhythmic action.

INDIRECT CLASS 4 ACTION

Acetylcholine

A brief stimulation of postganglionic vagal fibers supplying the SAN causes a large hyperpolarization for about 800 ms, followed by a short depolarization. In the surrounding transitional cells the hyperpolarization is less marked but longer lasting, and there is no subsequent depolarization. If acetylcholinesterase is inhibited, the duration of the effect is prolonged much more in the central than peripheral region, with the implication that the concentration of cholinesterase is highest in the center of the node (23). The hyperpolarization is caused by outflow of potassium through K_{Ach} channels. In the peripheral cells the hyperpolarization may provoke a failure of conduction ("exit block") from the dominant group of pacemaking cells, and the dominance may suddenly "jump" in the presence of acetylcholine (Ach) to a neighboring pacemaker (24). In the peripheral atrial cells APD is shortened, and the negative inotropic action of Ach is due to the failure of L-type calcium channels to remain open in response to the brief depolarization (see Chapter 3, Ref. 6).

The effect of Ach in this respect is to inhibit the influx of calcium, i.e., it has an indirect class 4 action.

The shortening of APD, and hence of ERP (Effective Refractory Period), increases the probability of reentry arrhythmia if there is an ectopic source of pacemaking. In canine heart-lung preparations electrical stimulation in the presence of cholinesterase inhibition (25) or of an infusion of Ach (26) invariably precipitated atrial fibrillation.

In the AVN there is a dense vagal innervation, with an action on N and N-H cells analogous to that on p cells and transitional cells in the SAN. Hyperpolarization of N cells opposes depolarization by incoming action potentials from the atrium, and hyperpolarization of N-H cells delays conduction and may induce heart block. Since calcium ions are the charge carriers in the AVN, AV block by cholinergic stimulation may again be defined as an indirect class 4 action.

There are Ach receptors on postganglionic sympathetic nerves, activation of which inhibits noradrenaline release. Thus, cholinergic stimulation has an indirect class 2 action also. If a cardioselective cholinergic agonist existed, it could be useful in the control of supraventricular arrhythmias.

Adenosine

Among the multiple actions of adenosine are some that resemble those of Ach, including vasodilatation via release of nitric oxide (27), and opening of K^+ channels in the AVN, which accounts for the efficacy of adenosine in supraventricular arrhythmias. Though less evanescent than Ach, adenosine has a very short life in blood, and is primarily useful for rapid control of the ventricular rate while other procedures are set in train. Adenosine also acts on the SAN, inducing bradycardia directly (28), markedly in isolated preparations, but in intact animals the effect may be overridden by reflex withdrawal of vagal input in response to hypotension. Adenosine activates A1 receptors, intracellular signaling involving guanosine triphosphate (GTP)-binding regulatory proteins coupled to potassium channels (29). In ventricular myocytes adenosine has an anti-sympathetic action also, inhibiting the adenosine $3',5'$-cyclic monophosphate (cAMP) cascade, but its antiarrhythmic action may be attributed to hyperpolarization and shortening of APD in the nodes, opposing depolarization by calcium ions, and it may therefore be categorized as an indirect class 4 action.

Digitalis

The fundamental action of cardiac glycosides on the heart is inhibition of the Na/K pump. In its therapeutic action on the whole body, however, an effect on the CNS is important, causing an increase in vagal activity, which slows the SAN frequency and delays AV conduction to protect the ventricle from a high rate. The central mechanism has been attributed to a sensitization of the baroreceptor reflex, either at the carotid sinus, or in the medulla, or both. The increased vagal activity is mainly responsible for the beneficial effects of low doses of cardiac glycosides in patients with intermittent atrial arrhythmias.

There are numerous active glycosides, digoxin being the most commonly prescribed in general practice, but for experimental studies ouabain is often preferred for its solubility. In isolated cardiac muscle the initial (30 min) effect of ouabain is to prolong APD, an effect attributed to inhibition of the electrogenic outward current supplied by the Na/K pump. In humans an initial lengthening of APD by cardiac glycosides has been revealed by monophasic action potential recording. A positive inotropic action is observed within a few minutes, which persists when the APD starts to shorten after an hour. At steady state APD is shorter than before drug administration.

In normal sinus rhythm the Na/K pump has reserve capacity, which is switched on when frequency rises, because sodium ions entering with each action potential activate the previously idle pumps, so that cytosolic $[Na^+]_i$ does not perceptibly increase. At the end of a rapid train of impulses the additional pumps are still active, and the cells briefly hyperpolarize,

at the same time as subsarcolemmal Na^+ decreases. Both these factors increase the inward driving force for Na^+, favoring Na/Ca exchange in the depolarizing direction (3 Na^+ in, 1 Ca^{2+} out), which accounts for the so-called triggered activity in the wake of a rapid train of action potentials.

When the Na/K pump is inhibited, several secondary effects are revealed. Cytosolic $[Na^+]_i$ increases, reducing the driving force for Na^+ entry, so that V_{max} is reduced. Subsarcolemmal Na^+ also increases, depressing Na/Ca exchange, so that calcium is extruded less rapidly and is captured by the SR. The amount of calcium released from the SR per beat is increased, causing a positive inotropy. APD eventually shortens, an effect that has been attributed to three mechanisms:

1. increased outward current through calcium activated K^+ channels;
2. accumulation of K^+ in the narrow intercellular clefts (30);
3. increased outward current through Na^+-activated K^+ channels.

K^+ current through these channels is decreased by class 3 antiarrhythmic drugs. In isolated guinea pig ventricular cells "amiodarone blocks the K^+_{Na} channels by inhibiting the openings in therapeutic concentrations. Ms-551 and E 4031 produce 'flickering' block in supratherapeutic concentrations" (31).

Cytosolic Ca^{2+} must fall below 10^{-7} M to permit muscular relaxation. In the presence of cardiac glycosides, the reduced Na/Ca exchange as subsarcolemmal Na^+ rises and the shortening of APD cause failure of adequate calcium extrusion until after the end of the action potential, when the negative resting potential favors the inward direction of exchange current. Consequently, the extrusion of Ca^{2+} by the depolarizing Na/Ca exchange extends into diastole, causing the transient inward current characteristic of digitalis intoxication. If these depolarizations reach threshold they are revealed as coupled extrasystoles. The final stage of intoxication is VT (Ventricular Tachycardia) or VF (Ventricular Fib), associated with a failure of relaxation as $[Ca^{2+}]_i$ rises.

The probability of cardiac glycosides inducing VT or VF is increased by several factors: (a) Rapid heart rate, as in atrial fibrillation: each action potential carries sodium into the cell, increasing the load on the Na/K pump. (b) A fall in plasma K^+, such as may occur in patients on diuretics. The Na/K pump cannot operate unless both external (K^+) and internal (Na^+) sites are occupied simultaneously. As $[K^+]_o$ falls the pump slows down, and in experimental situations ceases altogether at $[K^+]_o < 1$ mM. Inhibition of the pump by glycosides is not due to competition for the external K^+ site because the $[K^+]_o$ at which the pump current, I_p, is half maximal is unaffected by dihydroouabain (32). (c) High background sympathetic activity increases heart rate and lowers plasma K^+. In animals deprived of adrenergic stimulation by sympathectomy and adrenalectomy even lethal doses of cardiac glycosides do not induce VF. Protection against ouabain intoxication is provided by β-blockade (see Chapter 1, Ref. 12). In view of the dangers of arrhythmia induction, patients on glycosides may receive less surveillance than is prudent, especially if they are also taking diuretics.

In addition to the vagally induced effects on the nodes, cardiac glycosides have a direct action. SR is absent from the nodes, so that the rise of intracellular calcium caused by the reduced extrusion is more marked. In single cells isolated from the rabbit AVN strophanthidin reduced L-type inward calcium current in response to depolarizing pulses from 232 ± 65 pA to 48 ± 26 pA, and prolonged exposure led to the occurrence of spontaneous depolarizations. It was concluded that "strophanthidin reduces I_{Ca-L} in the AVN by an indirect effect, probably mediated by the rise in $[Ca^{2+}]_i$" (33). Thus, the beneficial effect of cardiac glycosides in protecting the ventricle from high rates by AV block can be attributed to an indirect Class 4 action.

General practitioners and cardiologists still prescribe cardiac glycosides for the treatment of atrial arrhythmias and congestive heart failure, but doubts have been expressed about their efficacy. The effect of digoxin on ven-

tricular rate can be overridden by adrenergic stimulation, and heart rate increases in exercise are not controlled (34). "Drugs commonly used for regulation of the ventricular rate are digitalis, calcium antagonists, and beta-blockers. Digitalis alone is not effective for adequate control of the ventricular rate during stress or exercise since it predominantly reduces the ventricular rate by potentiating the vagal tone. Sympathetic activation, which is also present in most haemodynamically compromised patients, will therefore overrule the effects of digoxin. Calcium antagonists (verapamil, diltiazem) and beta-blockers have a direct depressant effect on the AV node, but exert negative inotropic effects" (35).

"The aim of treating the symptomatic patient with atrial fibrillation is either to control the ventricular rate or to terminate the fibrillation. A randomized double-blind trial with digoxin, which aimed to convert recent-onset atrial fibrillation to sinus rhythm, clearly showed that digoxin was not superior to placebo in these patients. Most evidence shows that digoxin does not prevent [nor] abbreviate atrial fibrillation. The difficult diagnosis of the cause of arrhythmia postoperatively is much easier if the patient does not receive digitalis. Electrolyte disturbances or a reduction of renal function may provoke arrhythmias in the patient treated with digoxin" (36).

In heart failure the Bowdich effect and the Starling effect are lost, and the inotropic response to adrenergic agonists is reduced. The positive inotropic action of digoxin, however, is retained.

In conclusion, if cardiac glycosides are to be considered as antiarrhythmic drugs at all, which seems doubtful, their mode of action is as indirect Class 4 agents.

REFERENCES

1. Lin EC, Cantiello H. A novel method to study the electrodynamic behaviour of actin filaments. Evidence for cable-like properties of actin. *Biophys J* 1993; 65: 1371–1378.
2. Tsien RW, Ellinor PT, Zhang J-F, Bezprozvanny I. Molecular biology of calcium channels and structural determinants of key functions. *J Cardiovasc Pharmacol* 1996; 27(suppl A):S4–S10.

3. Uehara A, Yasukochi M, Imanaga I. Calcium modulation of single SR potassium channel currents in heart muscle. *J Mol Cell Cardiol* 1994; 26:195–202.

4. Hüsser J, Lipsius SL, Blatter LA. Calcium gradients during excitation-contraction coupling in cat atrial myocytes. *J Physiol* 1996; 494:641–651.

5. Trafford AW, Díaz ME, O'Neill SC, Eisner DA. Comparison of subsarcolemmal and bulk calcium concentration during spontaneous calcium release in rat ventricular myocytes. *J Physiol* 1995; 488:577–586.

6. Sipido KR, Carmeliet E, Pappano A. Na^+ current and Ca^{2+} release from the sarcoplasmic reticulum during action potentials in guinea-pig ventricular myocytes. *J Physiol* 1995; 489:1–18.

7. Evans AM, Cannell MB. Calcium influx through L-type calcium channels provides the major trigger for calcium-induced calcium release in guinea-pig ventricular myocytes. *J Physiol* 1995; 483P:14P–15P.

8. Negretti N, Varro A, Eisner A. Estimate of net calcium flux and sarcoplasmic reticulum calcium content during systole in rat ventricular myocytes. *J Physiol* 1995; 486: 581–591.

9. Wasserstrom JA, Vites A-M. The role of Na^+-Ca^{2+} exchange in activation of excitation-contraction coupling in rat ventricular myocytes. *J Physiol* 1996; 493(2):529–542.

10. Levi AJ, Issberner J. Effect on the fura-2 transient of rapidly blocking Ca^{2-} channel in electrically stimulated rabbit heart cells. *J Physiol* 1996; 493:19–38.

11. Levi AJ, Li J, Spitzer KW, Bridge HB. Effect on the Indo-1 transient of applying Ca^{2+} channel blocker for a single beat in voltage-clamped guinea-pig cardiac myocytes. *J Physiol* 1996; 494:653–673.

12. Vaughan Williams EM. A study of intracellular potentials and contractions in atria, including evidence for an after potential. *J Physiol* 1959; 149:78–92.

13. Harrison SM, Boyett MR. The role of Na^+-Ca^{2+} exchange in the rate-dependent increase in contraction in guinea-pig ventricular myocytes. *J Physiol* 1995; 482:555–566.

14. Ballard C, Schaffer S. Stimulation of the Na^+/Ca^{2+} exchanger by phenylephrine, angiotensin II and endothelin I. *J Mol Cell Cardiol* 1996; 28:11–18.

15. Wattano T, Kimura J, Mirita T Nakanishi H. A novel antagonist, No.7943, of the Na^+/Ca^{2+} exchange current in guinea-pig cardiac ventricular cells. *Br J Pharmacol* 1996; 119:555–563.

16. Fleckenstein A, Döring HJ, Kammermeier H. Experimental heart failure due to inhibition of utilization of high-energy phosphate. In: Marchetti G, Taccardi B, eds. *Symposium on the coronary circulation and energetics of the myocardium*. Basel: Karger, 1966; 220–236.

17. Vaughan Williams EM. Classification of antiarrhythmic actions. In: Vaughan Williams EM, ed. *Handbook of experimental pharmacology*. Heidelberg: Springer-Verlag, 1989; 45–67.

18. Melville KI, Shister AE, Huq S. Iproveratril: experimental data on coronary dilatation and antiarrhythmic action. *Can Med Assoc J* 1964; 90:761–770.

19. Munk AA, Adjemian RA, Zhao J, Ogbaghebriel A, Shrier H Electrophysiological properties of morphologically distinct cells isolated from the rabbit atrioventricular node. *J Physiol* 1996; 493:801–818.

20. Tamargo J, López-Sendón J. Relevance of the pharmacokinetic profile in the treatment of angina pectoris: the need for 24 h control. *Eur Heart J* 1995; 16:57–61.

21. Triggle DJ. The classification of calcium antagonists. *J Cardiovasc Pharmacol* 1996; 27(suppl A):S11–S16.

22. Reid JL, Petrie JR, Glen SK, Meredith PR, Elliott HL. Clinical pharmacology of the novel calcium antagonist mibefradil. *J Cardiovasc Pharmacol* 1996; 27(suppl A): S22–S26.

23. Kodama I, Boyett MR, Suzuki R, Honjo H, Toyama J. Regional differences in the response of the isolated sinus node of the rabbit to vagal stimulation. *J Physiol* 1996; 495:785–801.

24. Vaughan Williams EM. Some observations concerning the mode of action of acetylcholine in isolated atria. *J Physiol* 1958; 140:337–346.

25. Burn JH, Vaughan Williams EM, Walker JM. The production of block and auricular fibrillation in the heart-lung preparation by inhibition of cholinesterase. *Br Heart J* 1955; 17:431–437.

26. Burn JH, Vaughan Williams EM, Walker JM. The effect of acetylcholine in the heart-lung preparation including the production of auricular fibrillation. *J Physiol* 1955; 128:277–293.

27. Sterin-Borda L, Echague AV, Leiros CP, Genaro A, Burda E. Endogenous nitric oxide signalling system and the cardiac muscarinic acetylcholine receptor-inotropic response. *Br J Pharmacol* 1995; 115:1525–1531.

28. Bellni FL, Wang J, Hintze TH. Adenosine causes bradycardia in pacing-induced cardiac failure. *Circulation* 1992; 85:1118–1124.

29. Behnke N, Müller W, Neumann J, Schmitz W, Scholz H, Stein B. Differential antagonism by 1,3 dipropylxanthine-8-cyclopentyl xanthine, and 9-chloro-2-(2-furanyl)-5,6-dihydro-1,2,4-triazolo (1,5-c) quinazolin-5-imine of the effects of adenosine derivatives in the presence of isoprenaline on contractile response and cyclic AMP content in cardiomyocytes. Evidence for the co-existence of the A_1- and A_2-adenosine receptors on cardiac myocytes. *J Pharmacol Exp Ther* 1990; 254:1017–1023.

30. Herzig H, Lüllman H. The effects of cardiac glycosides at cellular level. In: Vaughan Williams EM, ed. *Handbook of experimental pharmacology*. Heidelberg: Springer-Verlag, 1989; 545–563.

31. Mori K, Saito T, Masuda Y, Nakaya H. Effects of class III antiarrhythmic drugs on the Na^+-activated K^+ channels in guinea-pig ventricular cells. *Br J Pharmacol* 1996; 119:133–141.

32. Hermans AN, Glitsch HG, Verdonck F. The antagonistic effect of K^+_o and dihydro-ouabain on the Na^+ pump current of single rat and guinea-pig cardiac cells. *J Physiol* 1995; 483:617–628.

33. Hancox JL. The digitalis analogue strophanthidin blocks L-type calcium current in single cells isolated from the rabbit atrioventricular node. *J Physiol* 1996; 493;25–26P.

34. Campbell RWF. Atrial fibrillation: steering a management course between thromboembolism and proarrhythmia risk. *Eur Heart J* 1995; 16(suppl G):25–31.

35. Lie KI, van Gelder IC. Therapy of recent-onset atrial fibrillation and flutter in haemodynamically compromised patients: chemical conversion or control of the ventricular rate? *Eur Heart J* 1995; 16:433–434.

36. Erdmann E. Digitalis—friend or foe? *Eur Heart J* 1995; 16(suppl F):16–19.

7

Class 5 Antiarrhythmic Action

E. M. Vaughan Williams

SPECIFIC BRADYCARDIC AGENTS

Alinidine

There are several approaches to antihypertensive therapy: direct vasodilatation, blockade of vasoconstrictor stimuli, reduction of cardiac output, or modulating vasomotor control by an action on the brain. Clonidine is an agonist at α_2-adrenoceptors, causing peripheral vasoconstriction, and was originally intended for use as a nasal decongestant. During clinical tests it was found to have activity on the CNS, including sedation and induction of bradycardia and hypotension by reducing the outflow of sympathetic impulses. Various analogues of clonidine were synthesized, one of which, the n-allyl derivative alinidine, induced a more profound bradycardia than clonidine (Fig. 1).

In healthy young male volunteers oral doses of 20, 40, and 80 mg of alinidine reduced resting heart rate (HR) and exercise tachycardia in a dose-related manner, significant reductions after 40 mg occurring at 1, 2, and 5 hours. Exercise HR was still reduced 6 hours after 80 mg from a mean of 150 beats·min^{-1} to 130 beats·min^{-1}, an effect similar to that of 40 mg propranolol. "Systolic and diastolic pressures in the standing position were not affected by 20 and 40 mg alinidine, but 80 mg reduced both, from 2 to 6 hours. Similar reductions were seen in the supine position after 80 mg alinidine. In all subjects 80 mg alinidine produced drowsiness, which was most pronounced about 8 to 9 hours after administration. 40 mg had no adverse effects" (1).

The dose of isoprenaline IV required to raise heart rate by 25 beats·min^{-1} was not altered by alinidine, from which it was concluded that the bradycardia was not related to blockade of β-adrenoceptors. The only metabolite of alinidine in humans is clonidine. A comparison of the effects of doses of clonidine producing plasma levels comparable with those occurring after the administration of alinidine led to the conclusion that the effects of alinidine on systolic and diastolic blood pressure were not caused by the clonidine metabolite. "The fall in supine and standing blood pressure is probably a direct effect of alinidine as it is greater at 2 hours than at 6 hours, whereas the plasma concentration of clonidine formed from alinidine is greater at 6 hours" (2). The appearance of the side effects of drowsiness and dry mouth, however, coincided in timing with the formation of the metabolite. During chronic administration of alinidine in 154 volunteers, drowsiness was reported in 65%, dry mouth in 44%, vertigo in 26%, headache in 28%, optical perseveration or blurring of vision in 23%, and hypotension in 13%, but "adverse effects disappeared after the first few days of treatment." Visual disturbance was reported in 26% of volunteers but in only 3.2% of patients. In early clinical trials, alinidine was "shown to be of benefit to patients with angina pectoris, reducing heart rate and nitrate consumption and increasing maximum symptom-free work load" (3).

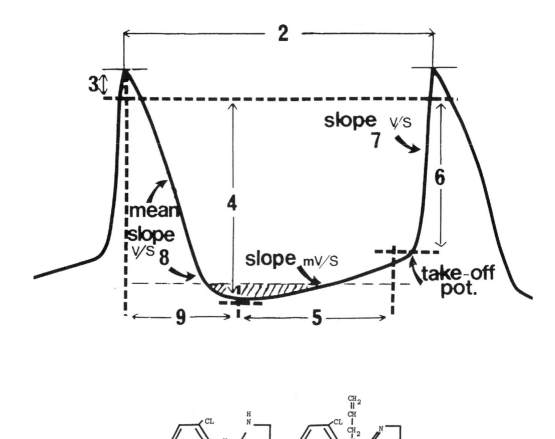

FIG. 1. Structures of clonidine and alinidine. *Top,* diagram of an intracellular potential of a transitional cell from the rabbit SA node: *2,* cycle length; *9,* repolarization time, from the peak to the maximum diastolic potential; *3,* overshoot potential, zero to peak; *4,* maximum diastolic potential (MDP); *1,* action potential amplitude (= 3 + 4); *7,* V_{max}, by differentiation of the upstroke; *8,* mean rate of repolarization, by differentiation of a tangent to the most linear portion of the repolarizing limb; *5,* slope of the slow diastolic depolarization, the mean slope of the middle third of the almost linear portion of the segment marked *5* between the MDP and the start of the upstroke; *6,* takeoff potential, obtained by extrapolating forward along *5,* and backward along *7.* All measurements were made by a computer program from digitized recorded potentials. The *shaded area* indicates the length of time the potential was negative to -50 mV, the activation threshold for the hyperpolarization-induced inward current, I_h.

Pharmacokinetics

After 40 mg alinidine IV the plasma concentration of alinidine fell rapidly from 921.7 ± 353.6 ng.ml^{-1} at the end of the infusion to 162.2 ± 49.3 ng.ml^{-1} after 1 hour. After 40 mg orally plasma levels rose from 4.7 ± 1.2 at 15 minutes to 120.1 ± 46.1 ng.ml^{-1} at 1 hour, reaching a maximum of 166.5 ± 18.5 ng.ml^{-1} at 1.8 ± 0.7 hours. The elimination half-life was 4.2 hours after both IV and oral doses. During chronic oral treatment with 40 mg bid peak and trough plasma concentrations remained stable from day 2 to day 8 at 356.2 ±

92.0 and 80.0 ± 35.8 ng.ml^{-1}, respectively. Thus, the therapeutic concentration of alinidine lies between 100 and 350 ng.ml^{-1} (0.3 to 1.0 μM), and in attempts to analyze the mode of action of alinidine, experimental studies employing concentrations higher than 2 μM would be invalid. (In a previous paper the plasma concentration of alinidine 1 hour after 40 mg had been reported to be 0.5 ± 0.2 ng.ml^{-1}, presumably a misprint for 0.5 μg.ml^{-1}.)

Mode of Action

In a detailed analysis of the electrophysiologic and pharmacologic effects of alinidine in various isolated cardiac preparations of the rabbit, the findings were mainly negative (see Chapter 1, Refs. 21–23). The major action of alinidine was to reduce the slope of the slow diastolic depolarization in sinus node cells (Fig. 1), a result confirmed by Haberl and Steinberg (4). Very high concentrations (29 μM) slightly prolonged APD, but only in transitional cells and atrial fibers. The maximum diastolic potential (MDP) in the dominant pacemakers and transitional cells was - 49 mV (4), more positive than the threshold for activation of the hyperpolarization-induced inward current, I_h, discussed later. Therapeutic concentrations of alinidine still caused a dose-related bradycardia in the presence of concentrations of cesium sufficient to block I_h (5).

The main findings were (a) alinidine is devoid of class 1 action, V_{max} is unchanged in all cardiac tissues, and QRS is unaltered; (b) alinidine does not block adrenoceptors; (c) therapeutic concentrations of alinidine had no effect on APD in atrial or ventricular muscle, and QT interval is unchanged in humans; (d) much higher concentrations (6.25 and 15.6 μM) increased APD by 8 and 15 ms, respectively, in rabbit atrial fibers; (e) alinidine had no negative inotropic action, nor did it alter the positive inotropic effect of raising extracellular calcium concentration; (f) the bradycardic action of alinidine was similar in hyperpolarizing (1.4 mM K^+_o) and depolariz-

ing (5.6 mM K^+_o) solutions; (g) the action was also unaffected by changes in pH (7 and 7.8) induced by CO_2/bicarbonate buffer, which alters intracellular pH also (see Chapter 1, Refs. 22–23).

Antiarrhythmic Action

In experiments on anesthetized dogs, arrhythmias were induced by administering IV doses of isoprenaline (0.2 μg.kg^{-1} + 0.4 + 0.4 . . . at intervals of 5 to 10 minutes) to animals breathing 1% halothane until the induction of multiple ectopic beats or VT. The arrhythmias were prevented by successive doses IV of alinidine 0.5 and 1.0 mg·kg^{-1} , the mean preventive dose being 2.9 ± 0.7 mg·kg^{-1}. Ouabain-induced arrhythmias were also abolished by alinidine, but much higher doses were required, up to 8 mg·kg^{-1}. Alinidine did not, however, protect against arrhythmias induced by coronary ligation (6).

Alinidine administered to patients undergoing cardiac surgery produced a plasma concentration of 172.5 ± 12 ng·ml^{-1} (0.49 μM) after 10 mg IV, and of 114 ± 13 ng·ml^{-1} (0.32 μM) after an infusion of 10 mg·h^{-1}. Stable heart rate reductions were achieved, without any effects on hemodynamics or ECG, and it was concluded that "alinidine appears to be a suitable drug for control of inappropriate sinus tachycardia in patients with heart disease undergoing surgery" (7). In a placebo-controlled study of the prophylactic use of alinidine in patients receiving coronary artery bypass grafts, 11 out of 16 controls suffered supraventricular tachycardias during the first 3 postoperative days, all of which required medical treatment. None of the alinidine-treated patients had sinoventricular tachyarrhythmia (SVT), and their mean heart rate was 82 ± 12 beats·min^{-1}, compared with 91 ± 21 beats·min^{-1} in controls (8).

Alinidine may be considered as an antiarrhythmic drug for supraventricular arrhythmias involving excessive sinoatrial node (SAN) activity, but it does not act by any of the four classes of action already established.

IONIC CURRENTS IN THE SINOATRIAL NODE

A cycle of transitional cell intracellularly recorded potentials is displayed in Fig. 1. Starting at the takeoff potential at -43 mV (6) the potential rises at a V_{max} of 13.6 V·s^{-1} (7) to a peak of +10 mV (3). The charge carrier is calcium flowing through L-type channels. I_{Ca-L} is increased by β_1-adrenoceptor, but not by β_2-adrenoceptor stimulation. From the peak, repolarization to the MDP of -53 mV (4) has a duration of 103 ms (9), and is produced by outward potassium current through delayed rectifier channels and by electrogenic Na/K pumping (9), (the Na$^+$ having entered, not through Na$^+$ channels, but via the Na/Ca exchange extruding calcium). The mean rate of repolarization is 0.62 V·s^{-1} (8), and is increased by β_1-adrenoceptor–stimulated K$^+$ conductance, and by β_2-adrenoceptor–stimulated Na/K pumping. The consequent shortening of APD accounts for about one-third of the reduction in cycle length (2) when heart rate is accelerated by sympathetic activity. Adrenergic stimulation also makes the MDP more negative.

In peripheral atrial fibers, and in ventricular and Purkinje cells, inward rectification channels open and lock the resting potential near to E_K. Inward rectification channels are absent from the SAN, so that as the outward current through the delayed rectifier channels declines, the potential reverses direction and a slow diastolic depolarization develops at a slope of 50 mV·s^{-1} (5), due to a background inward current. Eventually the potential is sufficiently positive to open L-type calcium channels at the takeoff potential, and the cycle is complete.

Two issues remain unresolved: (a) the charge carriers of the background current; and (b) any other current, or currents, that may be activated to increase the slope of the slow diastolic depolarization.

Part of the current depolarizing transitional cells must be passive "electrotonic" flow via gap junctions and extracellular clefts into the "sink" of the p cells, which are always positive to the transitional cells during diastole (10). Transitional cells, on reaching takeoff potential, depolarize more rapidly than p cells, however, and reach their peak before the p cells, so that for a few milliseconds the direction of current flow is reversed (Fig. 2). The dominant pacemakers and surrounding cells have been mapped, on the basis that the dominant cell is the first to reach a point halfway between MDP and peak potential (11). In the dominant cells selected on this basis there is no sharp transition between the slow diastolic depolarization and the upstroke, so that a sharp takeoff potential cannot be identified. Thus, depolarization spreads from the transitional cell peak in both directions, centripetally to accelerate depolarization of the p cell and centrifugally toward the rest of the atrium, and the true initiation of the heart beat occurs at the transitional/p cell junction (12).

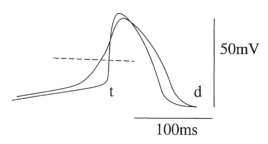

50mV

t   d

100ms

FIG. 2. Action potential from a transitional cell *(t)* superimposed on that of the central dominant pacemaker cell *(d)*, with the halfway point of the latter between MDP and peak *(dashed line)* occurring 12 ms before the halfway point of the *(t)* cell. The peak of the *(t)* cell is nevertheless reached earlier than that of the *(d)* cell.

Chloride Current

In 1963 de Mello (13) concluded that chloride ions contributed substantially to depolarizing current in the SA node: "When the membrane is repolarized, and the membrane potential exceeds [i.e., is negative to] E_{Cl} (-37 mV) , there is an outward movement of Cl⁻ ions which contributes to diastolic depolarization." Seyama (14) observed that in SAN cells "9% of the total membrane conductance of the resting potential is due to the Cl⁻ movement." In voltage clamp experiments he substituted various anions for chloride and observed the permeability sequence thiocyanate $> NO_3 > I > Br > Cl >$ acetate. The membrane resistance was measured, and "the conductance sequence for the sinoatrial node cell membrane was observed to be the same as the permeability sequence.... When organic anions were substituted for Cl⁻, the membrane gradually hyperpolarized; on readmission of Cl⁻ solution the membrane depolarized to the original potential." It was concluded that the Na/K pump current is normally short-circuited by chloride current:

> Hyperpolarization in acetate solution is due to electrogenicity produced by the Na-K pump.... In the sinoatrial node cell there is a current system showing inward going rectification and this current disappears after the administration of acetate solution. The shift of membrane potential to a more negative potential can be explained by the elimination of this current, so that the slow diastolic depolarization may be generated partly by this current (15).

In view of this evidence that an outflow of chloride ions contributed to the slow diastolic depolarization, the bradycardic action of alinidine on spontaneously beating rabbit atria was studied in the presence of solutions in which chloride had been replaced by less permeant, or more permeant anions. Bromide was chosen as the more permeant anion because in solutions of the even more permeant nitrate sinus rhythm became unstable. Methyl sulfate was the less permeant anion.

Substitution by methyl sulfate caused the spontaneous frequency to fall by a mean of 15.48 ± 4.34 beats·min⁻¹. In this solution the bradycardia produced by alinidine was much smaller than that seen in chloride ($p < .001$). The regression of the log-dose response curve for alinidine in methyl sulfate was not statistically significant ($p = .069$). In contrast, substitution of chloride by bromide significantly increased the spontaneous frequency by 6.0 ± 1.3 beats·min⁻¹ ($p = .002$) and its depression by alinidine was greater than in chloride. The observed effects do not prove, of course, that alinidine acts on anion-selective channels, but the results were at least consistent with the hypothesis that alinidine might restrict anionic current (see Chapter 1).

Chloride Current in Cardiac Myocytes

Isoprenaline 1 µM activated a time-independent chloride current in rabbit and guinea pig cardiac myocytes, adenosine 3',5'-cyclic monophosphate (cAMP) dependent, and insensitive to the K⁺ channel blocker 3,4-diaminopyridine (16). Excitation of α-adrenoceptors inhibited the chloride current induced by β-adrenoceptor stimulation, but not by forskolin, which causes a rise in cAMP by an intracellular mechanism (17). Lowering osmotic pressure induces a time-independent chloride current, but has no effect on the time-independent current induced by forskolin (18). A cAMP-independent increased chloride current is activated by the tyrosine-kinase inhibitor genistein. This Cl⁻ current is additive to the Cl⁻ current first maximally activated by forskolin (19). Cystic fibrosis patients have a genetic defect that causes failure of the cystic fibrosis transmembrane-conductance regulator gene (CFTR), which encodes a chloride channel. CFTR controls the cAMP-dependent Cl⁻ current in the heart (20). Angiotensin II also activates chloride current in rabbit atrial myocytes (21), and extracellular adenosine triphosphate (ATP), acting on P2 receptors, activates a chloride current in murine cardiac myocytes, independently of cAMP and calcium (22). In guinea pig cardiac myocytes chloride conductance is increased by phorbol

12-myristate-13-acetate (MPA), which activates protein kinase C (forskolin activates protein kinase A) (23).

The above evidence indicates that chloride conductance can be regulated by several different mechanisms, revealed in isolated cardiac myocytes from various species, though the function of such currents in the normal physiology of a beating heart is unclear. In atrial and SAN cells of the rabbit a time-independent chloride current, not influenced by cAMP or calcium, is activated by inflating the cells by a pressure of 5 to 15 cm H_2O (24).

In an attempt to identify the charge carrier of the background current, Hagiwara et al. (24) isolated rabbit SAN cells by dissection and treatment with elastase and collagenase. These cells would have had numerous gap junctions connecting them, which were assumed to have been sealed. Calcium injected into cardiac cells reduces intercellular conductance (25), and perhaps calcium enters gap junctions from the exterior, to seal them when the cells are separated. The isolated cells were studied by the whole-cell patch-clamp technique and subjected to the following procedures. $BaCl_2$ 1 mM and CsCl 2 mM were added to block the delayed rectifier K^- channels. $NiCl_2$ 2 mM was added to block the Na/Ca exchange as well as Ca^{2+} current. Ouabain 10 μM was added to block the Na-K pump. The hyperpolarization-activated current was blocked by superfusing with K^+-free solution containing 2 mM Cs^+. In several experiments 2 μM nifedipine was used to block Ca^{2+} channels.

Removal of Cl^- ions from the external solution or the addition of the Cl^- channel blocker DIDS failed to affect the membrane conductance. The current was proportional to the extracellular concentration of Na^+ over the range 0 to 200 mM. The current amplitude decreased in the order Rb > K > Cs > Na > Li, which suggests a poor cation selectivity of this current system (26).

"Under physiologic conditions, the reversal potential of the background current was around -21 mV at physiologic concentrations of intracellular K^+ and extracellular Na^{+}" (26).

Whether such experiments invalidate the conclusions of de Mello and Seyama is questionable. The latter had shown that the anionic current in the SAN was blocked by the simple addition of acetate, and in the presence of the complex cocktail of inhibitors described above the normal function of anion channels may have been disturbed also.

Hyperpolarization-Induced Current

An inward current activated by hyperpolarization (I_h) in small isolated strips of rabbit SA node (0.2 × 1 mm) was described by Noma et al. (27) and by Yanagihara and Irisawa (28). A similar current was later named I_f (29), but the term I_h is preferable to a funny current, both on grounds of priority and description of activation. The threshold voltage for activation is about -50 mV, and even at -70 mV activation was very slow, with a time constant of 2 to 4 seconds. It was concluded that since the maximum diastolic potential in SAN cells is -40 to -70 mV "in this voltage range I_h has too long a time constant to show dynamic changes during each cardiac cycle. It is unlikely that I_h has a significant effect on the normal action potential pattern" (28).

In Fig. 1 the time the diastolic potential is negative to the threshold for activation of I_h in a transitional cell is indicated by shading. In the dominant pacemaker cells MDP is even more positive; 1 to 2 mM of cesium completely blocks I_h at -70 mV, and blocks 90% at -90 mV. In studies of regional differences in the rabbit SAN Nikmaran et al. (30) observed that the maximum diastolic potential increases from <-50 mV in the central dominant region to >-60 mV in the transitional more peripheral cells. "Cs^+ 2 mM reduced the slope of the slow diastolic depolarization at the periphery of the node, but not in the center" (30). Even in cells with MDP sufficiently negative to be affected, 2 mM Cs "slowed but did not arrest spontaneous pacemaking. These experiments provide strong, though indirect, evidence for the presence in the SA node of a time-independent background current" (31).

Zaza et al. (32), employing the whole-cell patch-clamp technique on isolated single cells from rabbit SAN, suggested that I_h was inhibited by acetylcholine (Ach). In contrast, Boyett et al. (33), who had already shown that Ach had no direct action on inward calcium current in contracting atrium, but inhibited I_{Ca-L} indirectly by shortening APD, found that increased outward K^+ current was the dominant action in the SAN also. "We conclude that I_{K-Ach} plays an important role in the chronotropic effects of both low and high concentrations of bath-applied acetylcholine and of vagal stimulation, and that I_f and I_{Ca} play little or no role" (33).

Bouman et al. (34) applied long-lasting hyperpolarizing pulses (1 s) to small clusters of cells from the rabbit SAN, clamping the voltage in steps, and observed that I_h was not activated until the voltage was more negative than -60 mV. Application of alinidine 3 to 80 μM shifted "the negative part of the I-V relation in the hyperpolarizing direction (by 15 mV at 80 μM)" (34). The clinical bradycardic effect of alinidine cannot be attributed to block of I_h on the basis of this evidence, because the concentrations used were from 7 to 100 times greater than the therapeutic range, and no I_h was activated until the clamp potential exceeded -60 mV, more negative than the MDP of SAN dominant cells. Clamp pulses of 1000 ms were employed, whereas even in transitional cells, diastolic potential is negative to -50 mV for less than a tenth of this period. Very high concentrations of alinidine may well block I_h in isolated preparations strongly hyperpolarized for long periods, but this cannot be its only, or even its principal, bradycardic action, which persists in the presence of 2 mM cesium.

One of the potentially important features of I_f is its Cl⁻ sensitivity. Seyama first drew attention to the possibility that part of the inward-going rectification in SA node strands might be due to a time- and voltage-dependent Cl⁻ current. Yanagihara and Irisawa observed a marked reduction of I_f when all the Cl⁻ in the Tyrode solution was removed and replaced with acetate or proprionate. One possible explanation of this decrease could be the change of intracellular pH when Cl⁻ is removed (35).

This explanation cannot apply to the reduction of the bradycardic effect of alinidine by the substitution of Cl⁻ with less permeant anions, because the bradycardic effect was the same when pH was changed to 7 or 7.8 by alterations in CO_2/bicarbonate buffer, which alters intracellular pH also (see Chapter 1, Ref. 22).

"It has now been shown that Cl⁻, while not acting as a charge carrier of I_f itself, exerts a modulatory effect on the mixed Na^+-K^+ conductance" (35). Even if the charge carriers of I_h are not Cl⁻ ions themselves "the conductance is unique in that it is dependent on an extracellular anion (Cl⁻)" (36). The possibility that the background current is carried by chloride is not excluded by these experiments. As to the effect of alinidine on I_h it is still possible that it acts as an anion antagonist by affecting the control exerted by chloride ions on the hyperpolarizing-induced current.

ZERO TIME OF THE PACEMAKING CYCLE

In Fig. 2 the peak of the transitional cell potential occurs before that of the dominant pacemaking cell, although the halfway point of the latter is reached earlier. In isochrone maps of the SAN and surrounding regions, the contours are more widely separated the further from the node (11). The action potential accelerates as it spreads into the atrium, because sodium current dominates more and more as inward rectification current makes the diastolic potential more negative, and the slow diastolic depolarization disappears. The cells with more negative resting potentials in which sodium current is activated are physically larger than cells without sodium current (37). "In the periphery I_{Na} plays an important role in pacemaking, whereas in the center it plays no role; in the center, but not in the periphery, I_{Ca} is obligatory for pacemaker activity" (38). Thus, the true zero, the "initiation" of the heart beat, occurs at the border between the dominant and transitional cells.

OTHER SPECIFIC BRADYCARDIC AGENTS

Alinidine was the prototype specific bradycardic drug, but several other specific bradycardic agents (SBA) have been synthesized, though their modes of action may differ (Fig. 3).

Falipamil

In a placebo-controlled study on 12 human volunteers, in the placebo group a 0.25 mg cumulative dose of atropine induced a 16% decrease in heart rate; a 1.0 mg cumulative dose increased the heart rate by 23%. After pretreatment with 100 mg falipamil heart rate decreased by 14% after low-dose atropine, while a 1.0 mg cumulative dose raised the heart rate by 3%. A 200 mg cumulative dose of falipamil resulted in a 20% decrease in heart rate after 0.125 + 0.125 mg atropine. A 1.0 mg dose of atropine was followed by a 8%

decrease in heart rate. There were no significant changes in blood pressure or ECG (39).

In a randomized study in patients with ischemic heart disease falipamil 2mg·kg^{-1} IV reduced heart rate on submaximal exercise from 129 ± 3 to 113 ± 3 beats·min^{-1}, without change of blood pressure. Falipamil also increased the time at which exercise was terminated by angina or dyspnea from 549 to 607 seconds (40).

Zatebradine

In a double-blind placebo-controlled study in 32 healthy male subjects, zatebradine 2.5, 5, or 10 mg was administered orally every 6 hours 3 times daily for 7 days. The mean values for resting heart rate from day +1 to day +7 were 70.9 on placebo, and 61.3, 59.4. and 55.1 beats·min^{-1} for the three treated groups respectively. The bradycardia persisted for the 11 hours between the last dose of one day and the first of the next, in spite of the fact

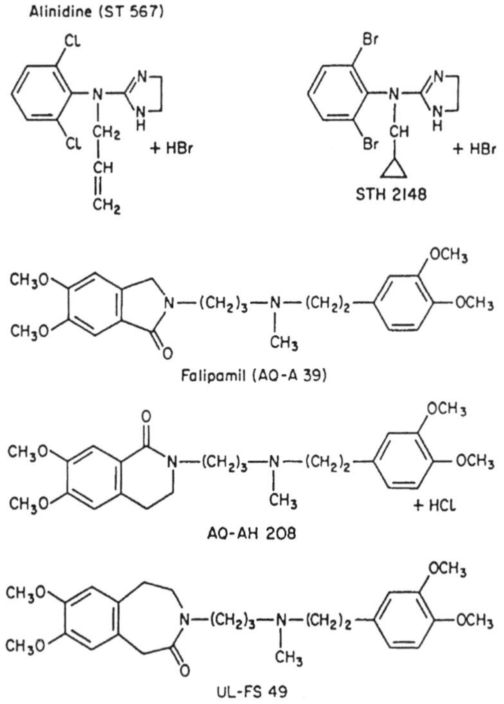

FIG. 3. Structures of specific bradycardic drugs. UL-FS 49 is now called zatebradine. Cl (m.w. 493).

that the plasma half-life of zatebradine is only 2 to 3 hours. No significant changes in blood pressure were observed, and the QT_C was unchanged on exercise. The heart rate during exercise was reduced by 10% to 20%. Visual disturbances were observed in six subjects on the highest dose (39).

Zatebradine blocked the hyperpolarization-induced current in isolated rabbit SAN cells, but its mode of action differed from that of alinidine. "Alinidine and its derivatives primarily shift the activation curve of I_f to more hyperpolarized potentials, whereas zatebradine and related compounds cause a use-dependent block of I_f without shifting the activation curve" (41). The concentration used in these experiments was said to be clinically relevant, but was, in fact, 1 μm or 493 ng·ml^{-1}. Even during treatment with the highest dose of zatebradine used in humans, 10 mg t.i.d., the maximum plasma concentration was only 35 ng·ml^{-1}, and the therapeutic range was 15 to 25 ng·ml^{-1}. The concentration of zatebradine required to block I_f was thus at least 20 times greater than the clinically relevant level.

The AV node, sometimes called the pacemaker of the left ventricle, contains N cells, analogous to p cells of the SAN, which have the capacity to pacemake spontaneously if released from atrial control, and N-H cells, analogous to transitional cells. In anesthetized dogs zatebradine 0.1 to 0.3 μmol·kg^{-1} IV reduced the frequency of subsidiary AVN pacemakers (42). In pigs, zatebradine was as effective as atenolol in reducing infarct size after coronary occlusion, and in the nonischemic zone contractile function was actually better after zatebradine than after atenolol (43).

SC 16257

SC 16257 is a new specific bradycardic agent that reduces sinus rate without any negative inotropic action, and that increases aortic flow, the stroke volume being augmented to compensate for the slower heart rate (44). SC 16257 exhibited a dose-related inhibition of I_h in rabbit SAN cells (IC$_{50}$ 2.8

μM). In contrast to zatebradine, which depressed I_K by 20%, 3 μM of SC 16257 had no effect on I_K (or on I_{Ca-T}). The activation curve of I_h was not shifted by SC 16257, but the magnitude of the fully activated current was reduced (45).

CONCLUSION

Specific bradycardic agents may primarily be of use in the treatment of angina pectoris and ischemic heart disease, but they have a unique and selective effect on SAN cells, and could be beneficial as antiarrhythmic drugs for the treatment of paroxysmal supraventricular tachycardia and for prophylaxis against arrhythmias following surgery. Very high concentrations block the hyperpolarization-induced inward current I_h, but in the experiments that have demonstrated this property in isolated preparations, concentrations 10 to 100 times greater than therapeutic levels were employed. Alinidine also induced bradycardia in the presence of Cs$^+$ 1 to 2 mM, which blocks I_h, suggesting that this effect cannot be its only action in the SAN.

Whatever the mode of action of SBA, it is certainly different from the other four classes of antiarrhythmic action already established, and merits categorization as a fifth class.

REFERENCES

1. Harron DWG, Riddell JG, Shanks RG. Alinidine reduces heart rate without blockade of beta-adrenoceptors. *Lancet* 1981; 1:351–353.
2. Shanks RG. The clinical pharmacology of alinidine and its side effects. *Eur Heart J* 1987; 8(suppl L):83–90.
3. Harron DNG, Shanks RG. Pharmacology, clinical pharmacology and potential therapeutic uses of the specific bradycardic agent, alinidine. *Eur Heart J* 1985; 6: 722–729.
4. Haberl R, Steinbeck G. Chronotropic action of alinidine on the isolated sinus node of the rabbit heart. *Eur Heart J* 1985; 6:730–736.
5. Dennis PD, Vaughan Williams EM. Further studies of alinidine-induced bradycardia in the presence of cesium. *Cardiovasc Res* 1986; 20:375–378.
6. Harron DWG, Allen JR, Wilson R, Shanks RG. Effect of alinidine on experimental cardiac arrhythmias. *J Cardiovasc Pharmacol* 1982; 4:221–225.

7. Skarvan K. Value of a specific bradycardic agent in cardiac surgery compared to placebo. *Eur Heart J* 1987; 8(suppl L):123–129.
8. Kleinpeter U-M, Iversen S, Tesch A, Schmiedt W, Mayer E, Oelert H. Prevention of supraventricular tachyarrhythmias post coronary artery bypass surgery. *Eur Heart J* 1987; 8(suppl L);137–140.
9. Sakai R, Hagiwara N, Matsuda N, Kasanuki H, Hosoda S. Sodium-potassium pump current in sino-atrial node cells. *J Physiol* 1996; 490:51–62.
10. Sohn HG, Vassalle M. Cesium effects on dual pacemaker mechanisms in guinea-pig sinoatrial node. *J Mol Cell Cardiol* 1995; 27:563–577.
11. Masson-Pévet M, Bleeker WK, Gros D. The plasma membrane of leading pacemaker cells in the rabbit sinus node. *Circ Res* 1979; 45:621–629.
12. Vaughan Williams EM. Cardiac electrophysiology. In: Vaughan Williams EM, ed. *Handbook of experimental pharmacology.* Heidelberg: Springer-Verlag, 1989; 1–44.
13. de Mello WC. Role of chloride ions in cardiac action and pacemaker potentials. *Am J Physiol* 1963; 205:567–575.
14. Seyama I. The effect of Na, K and Cl ions on the resting membrane potential of sinoatrial node cell of the rabbit. *Jpn J Physiol* 1977; 27 577–588.
15. Seyama I. Characteristics of the anion channel in the sino-atrial node of the rabbit. *J Physiol* 1979; 294: 447–460.
16. Harvey RD, Hume JR. Isoproterenol activates a chloride current, not the transient outward current, in rabbit ventricular myocytes. *Am J Physiol* 1989; 217:C1177–1181.
17. Iyadomi I, Hirahara K, Ehara T. (α-Adrenergic inhibition of the β-adrenoceptor-dependent chloride current in guinea-pig ventricular myocytes. *J Physiol* 1995; 489:95–104.
18. Shuba LM, Ogura T, McDonald TF. Kinetic evidence distinguishing volume sensitive chloride current from other types in guinea pig ventricular myocytes. *J Physiol* 1996; 491:69–80.
19. Shuba LM, Asai T, Pelzer S, McDonald TF. Activation of cardiac chloride conductance by the tyrosine kinase inhibitor, genistein. *Br J Pharmacol* 1996; 119:335–345.
20. Warth JD, Hart P, Horowitz B, Hume JR. CFTR antisense oligodeoxynucleotides reduce cAMP-dependent Cl$^-$ currents in cultured guinea-pig ventricular myocytes. *J Physiol* 1995; 487:128P.
21. Morita H, Kimura J, Endoh M. Angiotensin II activation of a chloride current in rabbit cardiac myocytes. *J Physiol* 1995; 483:119–130.
22. Levesque PC, Hume JR. ATP$_0$ but not cAMP$_i$ activates chloride current in mouse ventricular myocytes. *Cardiovasc Res* 1995; 29:336–343.
23. Shuba LM, Asai T, McDonald TF. Phorbol ester activation of chloride current in guinea-pig ventricular myocytes. *Br J Pharmacol* 1996; 117:1395–1444.
24. Hagiwara N, Masuda H, Shoda M, Irisawa H. Stretch-activated anion currents of rabbit cardiac myocytes. *J Physiol* 1992; 456:285–302.
25. de Mello WC. Effect of intracellular injection of calcium and strontium on cell communication in the heart. *J Physiol* 1975; 250:231–245.
26. Hagiwara N, Irisawa H, Kasanuki H, Hosoda S. Background current in sinoatrial node cells of the rabbit heart. *Jpn J Physiol* 1992; 448:53–72.
27. Noma A, Yanagihara K, Irisawa H. Inward current of the rabbit sinoatrial node cell. *J Physiol* 1977; 372:43–51.
28. Yanagihara K, Irisawa H. Inward current activated during hyperpolarization in the rabbit sinoatrial node. *Pflugers Arch* 1980:385:11–19.
29. Brown HF, DiFrancesco D, Noble SJ. How does adrenaline accelerate the heart? *Nature* 1979 280; 235–236.
30. Nikmaran MR, Boyett MR, Kodama I, Suzuki R. Regional differences in the role of the hyperpolarization-activated current, i$_f$, in the pacemaker potential in the isolated sinoatrial node of the rabbit. *J Physiol* 1995; 487:126P–127P.
31. Denyer JC, Brown HF. Pacemaking in rabbit isolated sino-atrial node cells during Cs$^+$ block of the hyperpolarization-activated current. *J Physiol* 1990; 429:401–409.
32. Zaza A, Robinson RB, Francesco D. Basal response of the L-type Ca^{2+} and hyperpolarization-activated currents to autonomic agonists in the rabbit sino-atrial node. *J Physiol* 1996; 491:347–356.
33. Boyett MR, Kodama I, Honjo H, Arai A, Suzuki R, Tuyama J. Ionic basis of the chronotropic effect of acetylcholine on the rabbit sinoatrial node. *Cardiovasc Res* 1995; 29:867–879.
34. Bouman LN, Jongsma HJ, Opthof T, van Ginneken ACG. Does I$_f$ contribute to pacemaking in the rabbit sinoatrial node? *J Physiol* 1985; 358:51P.
35. Irisawa H, Brown HF, Giles W. Cardiac pacemaking in the sinoatrial node. *Physiol Rev* 1993; 73:197–227.
36. Frace AM, Maruoka F, Noma A. Control of the hyperpolarization-activated current by external anions in rabbit sino-atrial cells. *J Physiol* 1992; 453:307–318.
37. Honjo H, Boyett MR, Kodama I, Toyama J. Correlation between electrical activity and the size of rabbit sino-atrial cells. *J Physiol* 1996; 496:795–808.
38. Nikmaram MR, Kodama I, Boyett MR, Suzuki R, Honjo H. The Na$^+$ current plays an important role in pacemaker activity in the periphery, but not in the centre, of the rabbit sinoatrial node. *J Physiol* 1996; 491: 154P.
39. Franke H, Su CAPF, Schumaker K, Seiberling M. Clinical pharmacology of two specific bradycardic agents. *Eur Heart J* 1987; 8(suppl L):91–98.
40. Gilfrich HJ, Oberhoffer M, Witzke J. Comparison of AQ-A 39 with propranolol and placebo in ischaemic heart disease. *Eur Heart J* 1987; 8(suppl L):147–151.
41. Goethals M, Raes A, van Bogaert P-P. Use dependent block of the pacemaker current I$_f$ in rabbit sinoatrial node cells by zatebradine (UL-FS 49). *Circulation* 1993; 88:2389–2401.
42. Yamazaki K, Furukawa Y, Nakano H, Kasama M, Imamure H, Chiba S. Inhibition of the subsidiary pacemaker activity by zatebradine, an I$_f$ inhibitor, in the anaesthetized dog heart. *J Cardiovasc Pharmacol* 1995; 26:957–964.
43. Schulz R, Rose J, Skyschally A, Heusch G. Bradycardic agent UL-FS 49 attenuates ischemic regional myocardial dysfunction, and reduces infarct size in swine: comparison with the β-blocker atenolol. *J Cardiovasc Pharmacol* 1993; 25:216–228.
44. Gardiner SM, Kemp PA, March JE, Bennett T. Acute and chronic cardiac and regional haemodynamic effects of the novel bradycardic agent, SC 16257, in conscious rats. *Br J Pharmacol* 1995; 115:579–586.
45. Bois P, Bescond J, Renaudon B, Lenfant J. Mode of action of the bradycardic agent SC16257 on ionic currents of rabbit sinoatrial node cells. *Br J Pharmacol* 1996; 118:1051–1057.

8

Risk Factors: Ischemia and Hypertrophy

E. M. Vaughan Williams

ISCHEMIC HEART DISEASE AND RISK OF ARRHYTHMIA

Cardiac hypoxia may be defined as a lowering of the oxygen saturation of the blood supplying the heart, without any reduction in blood flow. Easily induced experimentally, it is rarely experienced in life except accidentally during anesthesia, or by ascent to high altitude. Ischemia involves restriction or interruption of blood flow, and is inevitably accompanied by hypoxia. In humans ischemia is usually slow in onset, and the blockage of an artery is habitually preceded by a gradual narrowing, providing an opportunity for the development of collateral flow. The choice of an animal model of ischemia to imitate the human situation is difficult, usually involving restriction of flow in a perfused heart or the production of an infarct by single or two-stage ligation of a coronary artery. Dogs, pigs, cats, ferrets, rabbits, guinea pigs, and rats have been used, the latter being the least appropriate because the ventricular action potential has no plateau, and the rapid heart rate and short diastole make the cardiac electrophysiology very different from that of man. Lack of oxygen induces progressive changes, ending in cell death, some of which may be accelerated and exaggerated by the consequences of reduced flow, such as accumulation of extracellular potassium and hydrogen ions.

Myocardial ischemia increases the risk of arrhythmia, and VT or VF is often the terminal event after an acute myocardial infarction (MI). In patients resuscitated out of hospital after collapse with VT or VF, recurrence of arrhythmia was more frequent in those with-

out overt MI than in those with proven MI. The implication is that several small lesions are more arrhythmogenic than an infarct large enough to release enzymes in measurable quantities. "Ventricular arrhythmias are a frequent finding in heart failure, and heart failure is a major underlying condition which is correlated to sudden death. Therefore, both sudden death and death from progression of heart failure strongly overlap" (1).

Many attempts have been made, in biochemical and electrophysiologic studies in a variety of animal models of ischemia, to identify the most important changes that could increase the probability of arrhythmia, in the hope that an appropriate intervention might reduce the risk.

"The role of ventricular arrhythmias and sudden death in patients with heart failure is still an enigma. Currently we should treat heart failure, but present data do not support antiarrhythmic drug treatment of asymptomatic complex arrhythmias detected on Holter monitoring. Given the poor survival rate of out-of-hospital cardiac arrest, the most important challenge will be to clearly identify and prophylactically treat potentially reversible or only transiently active arrhythmogenic factors" (1).

PROGRESSIVE EFFECTS OF HYPOXIA AND ISCHEMIA

Exposure of spontaneously beating isolated cardiac preparations to mild hypoxia (20% O_2) with full irrigation by glucose-containing medium causes an immediate reduction in contractions and a shortening of APD. There is no significant fall of adenosine triphosphate (ATP) or change in pH, and magnetic resonance studies have shown that the negative inotropic effect is associated with a rise in inorganic phosphate. The phosphate may restrict the release of calcium from the sarcoplasmic reticulum (SR), or interfere with the attachment of calcium to troponin. The shortening of APD would itself have a negative inotropic action by reducing calcium entry. ATP-sensitive K^+ channels do not open unless $[ATP]_I$ falls to micromolar levels, so that if the APD-shortening is attributed to increased I_{K-ATP}, it must be assumed that the hypoxia causes a local subsarcolemmal ATP deficiency, isolated from the millimolar levels of cytosolic ATP. The shorter APD is associated with a shorter ERP, increasing the probability of reentry in the presence of an ectopic source.

With more severe and prolonged hypoxia/ischemia, lactate production causes a fall in pH, and contraction decreases further as ATP and creatine phosphate (CrP) are consumed. The fall in pH induces a prolongation of the tail of the action potential (2), and depression of Na/K pumping increases Na^+_i, which decreases V_{max}. Whether these latter effects are likely to be antiarrhythmic or arrhythmogenic is debatable. The slowing of conduction velocity disrupts the normal sequence of excitation, but the prolongation of APD increases ERP. Both the accumulation of Na^+_I and the less negative resting potential reduce Na/Ca exchange and $[Ca]_i$ rises, leading to calcium overload and eventually to failure of relaxation and contracture.

In acute total ischemia in sheep cardiac Purkinje fibers and guinea pig papillary muscles induced by immersing them in paraffin oil, extracellular K^+ increased, and pH fell both inside and outside the cells. APD was shortened, but this was not due to the pH change, because reduction of pH without ischemia prolonged APD (in confirmation of reference 2). The shortening of APD was blocked by glibenclamide (3). Woodcock et al. (4) observed that after global ischemia inositol (1,4,5)triphosphate [Ins (1,4,5) P_3] was released, and suggested that this com-

pound might be involved in the development of ischemia-induced arrhythmias. Gentamicin reduced the formation of labeled [^{3}H] Ins (1,4,5) P_3, without any effect on other inositol phosphates, and reduced the incidence of the arrhythmias.

In globally ischemic heart models cell death occurs after 15 to 20 minutes, causing release of enzymes and troponin into the plasma. If the ischemia is localized, e.g., by ligation of a small coronary artery, the depolarized region acts as a sink into which current flows from the surrounding normoxic region. Arrhythmias arise from the border region neighboring an experimental infarct. The presence of a nonconducting region disrupts the normal sequence of excitation, encouraging reentry.

ROLE OF THE SYMPATHETIC NERVOUS SYSTEM

In addition to direct effects on myocardial function, ischemia has secondary effects on the CNS. Pain may be severe, and there is a massive release of noradrenaline from reflexively activated sympathetic nerves. α_1, α_2, β_1, and β_2-adrenoceptors mediate their individual actions (see Chapter 3), but adrenergic stimulation also modulates intracellular pH in the heart. The Na^+/H^+ exchange and the Na^+/HCO_3^- cotransport systems are alkalinizing, and the Cl^-/HCO_3^- exchange is acidifying. α_{1a}-Adrenoceptor stimulation raises intracellular pH in atrial and cardiac Purkinje cells via Na^+/H^+ exchange, blocked by amiloride and by prazosin, and noradrenaline dose-dependently increased $[pH]_i$ in rat ventricular myocytes. Stimulation of purinoceptors by ATP (released with noradrenaline from sympathetic varicosities) also raises $[pH]_i$. In contrast, selective β_1-adrenoceptor stimulation inhibited Na^+/H^+ exchange, slowing recovery from acidosis and lowering $[pH]_i$ (5). Thus, in spite of the alkalinizing effect of activation of α_1-adrenoceptors, the overall result from the large noradrenaline release exacerbates the effects of hypoxia, increasing calcium entry and shortening APD further.

In addition to the reflex activation of sympathetic nerves, noradrenaline release may be modulated directly by hypoxia. Presynaptic α_2-adrenoceptors activated by noradrenaline induce negative feedback, inhibiting further noradrenaline release. This effect is lost in hypoxia, but the facilitatory action of β_2-adrenoceptors and of angiotensin II is not. Exocytosis is also increased by hypoxia, the local direct effects of which exacerbate the excessive noradrenaline release in ischemic tissue (6).

APD-shortening occurs early in mild hypoxia, when ATP is not depleted, suggesting the possibility that a repolarizing current other than I_{K-ATP} was activated. Petrich et al. (7) concluded, from experiments on isolated rabbit hearts subjected to hypoxia, that the early APD-shortening was caused by chloride current activated by stimulation of β-adrenoceptors by endogenously released noradrenaline. The hypoxic APD-shortening was abolished by prior treatment with reserpine and by nadolol, and by lowering chloride concentration. "The inability of glibenclamide to block early APD reduction truly reflects a negligible role of I_{K-ATP} at this stage" (7).

Sympathectomy or prolonged beta-blockade causes an adaptive prolongation of cardiac APD (see Chapter 3). Sympathetic nerves are destroyed within an infarcted region, so that "sympathetic denervation occurs distal to the area of necrosis and delays repolarization." Agetsuma et al. (8), commenting on the transient giant negative T-wave (GNT) that may be observed after an acute anterior MI, suggested "that another possible mechanism of GNT may be the presence of a denervated but viable myocardium with delayed repolarization, over the endocardial myocardium with shortened repolarization."

The angiotensin-converting enzyme (ACE) inhibitors, originally introduced for the treatment of hypertension, have proved unexpectedly beneficial in patients with congestive heart failure, for reasons not fully understood, but involving more than one neurohormonal modulation (9). ACE inhibitors have many actions, one of which is to modulate noradrenaline release from sympathetic nerves. Microelectrode studies on guinea pig ventricular and sinoatrial node (SAN) preparations showed that captopril had no effect on V_{max}, either in control solutions or in simulated ischemia ($[K^+]_o$ 11.2 mM, $[pH]_o$ 6.4, pO_2 <80 mm Hg), and did not increase the normal small frequency-dependent reduction of V_{max}. There was no effect on SAN function (10). Ace inhibitors, like beta-blockers, reduce mortality in post-MI patients. Experimentally, enalapril and captopril reduced the incidence of reperfusion-induced arrhythmias. In patients with suspected MI, captopril reduced the incidence of ventricular ectopic beats (VEB) from $0.84 \cdot h^{-1}$ in patients on placebo to $0.48 \cdot h^{-1}$ on the third day, and from $0.77 \cdot h^{-1}$ to $0.51 \cdot h^{-1}$ after 2 weeks (11). The antiarrhythmic action of ACE-inhibitors may, at least in part, be attributed to an indirect class 2 action, by reducing sympathetic activity and blocking the adrenergic-enhancing effect of angiotensin II.

In conclusion, the sympathetic system is involved in ischemia, not only through the massive release of noradrenaline acutely, but also long-term by adaptive changes resulting from sympathetic denervation within and beyond a necrotic zone, which can disrupt the normal pattern of repolarization in the surviving myocardium.

SODIUM CURRENT IN HYPOXIA/ISCHEMIA

Initially V_{max} is hardly affected by hypoxia, but it is eventually depressed when the sodium pump begins to fail and $[Na^+]_i$ rises and resting potential becomes less negative. This creates a disparity between a slower conduction velocity in the ischemic zone and the normal velocity in surrounding tissue. Class 1 agents can be beneficial in acute local ischemia because (a) conduction velocity is depressed in the normoxic myocardium, reducing the disparity; (b) electric threshold is raised in the nonischemic regions, reducing the probability of invasion by an ectopic impulse originating in the border zone; and (c) V_{max} is depressed more in the partially depolarized ischemic zone, which may become electrically silent.

The therapeutic concentration of class 1b compounds is 20 to 100 times greater than that of class 1c drugs. After depolarization most Na channels are rapidly blocked by class 1b drugs, so that none is available at the beginning of diastole and ERP is prolonged. After repolarization, class 1b drugs are rapidly released, so that in sinus rhythm by the end of diastole all channels may again be drug free. If, however, the resting membrane potential is less negative, the class 1b compounds may not be released, which explains why lidocaine and mexiletine exert a greater effect on ischemic than on normal tissue. In rat ventricular myocytes the persistent (noninactivating, or late activating) plateau sodium current was potentiated by hypoxia, but both this current and the transient sodium current were abolished by lidocaine and tetrodotoxin (TTX) (12). With class 1c compounds, with a low therapeutic concentration, only a minor proportion of sodium channels are blocked; the remainder, being drug-free, recover at the normal time, so that if sufficient are available to produce a conducted impulse at all, ERP is not prolonged. If, however, the condition of the myocardium deteriorates, more and more of the previously free channels may take up the drug, and a vicious circle ensues, leading ultimately to cardiac arrest. Cibenzoline has a property additional to its class 1 action, in reducing the APD-shortening response to hypoxia (see Chapter 2).

In post-MI patients, class 1 agents do not reduce mortality, and class 1c compounds may increase it. Some authors assumed that the increased mortality in CAST (Cardiac Arrhythmia Suppression Trial) (see Chapter 2) could be attributed to a proarrhythmic effect of class 1c drugs, but firm evidence was lacking because many patients died without the immediate cause of death being established. In progressive ischemic heart disease the combination of renewed ischemia and further depression of V_{max} by unmodified admin-

istration of class 1c drugs could lead to conduction failure and cardiac arrest. Greenberg et al. (13), after a fresh analysis of the CAST data, argued that the drug treatment could have converted a renewed ischemic episode, nonfatal in the untreated group, into a fatal failure in the treated group.

Evidence obtained from studies of rat ventricular myocytes has suggested that hypoxia may induce an increase in the persistent sodium current. This may flow through late-opening or late-inactivating channels, but in the rat the ionic currents differ substantially from those of rabbits, dogs, or humans, because the rat ventricular action potential has no plateau. The authors suggest that "it is possible that hypoxia causes activation of [protein kinase C] PKC, which then modifies the structure of Na^+ channels, and induces resistance to inactivation" (14). The implication is that persistent inward sodium current could be arrhythmogenic, but experiments on isolated rat ventricular myocytes may not be relevant to the effects of hypoxia/ischemia in humans.

CALCIUM CURRENT IN HYPOXIA/ISCHEMIA

Shortening of APD in the early stage of hypoxia reduces inward calcium current (indirect class 4 action, analogous to the effect of Ach). Eventually, when stores of CrP and glycogen are depleted in spite of reduced consumption by depressed contractions, the Na pump cannot keep pace with the load, and $[Na-]_i$ rises. The driving force for calcium extrusion is reduced and the Na/Ca exchange may go into reverse during the action potential plateau. Consequently $[Ca^{2+}]_i$ rises, until the muscle can no longer relax fully (calcium overload), and a contracture develops. In experimental studies of total ischemia in whole hearts applied for 15 to 20 minutes, after readmission of oxygen the status quo is restored very slowly, because so many functions are competing. Stores of HE (High Energy) phosphate must be replaced, lactate oxidized, and glycogen replenished. Contractions take a long time to recover fully (the heart is "stunned"). In the immediate reperfusion period arrhythmias are common, and have more than one precipitating cause. Free radicals may be formed, causing disruption of intracellular membranes and release of enzymes. When contractions restart, cells may lose their preischemic geometry, and induce leakage through gap junctions (15). Renewed Na/Ca exchange may continue into diastole, and since extrusion of calcium is depolarizing transient inward current may reach threshold and induce an extrasystole.

Calcium overload can be induced experimentally by conditions that could not occur in life, such as poisoning with Cs^+ and cardiac glycosides, and exposure to an abnormal ionic environment. In such circumstances uptake and release of Ca^{2+} from the SR becomes unstable, and oscillatory rises and falls of $[Ca^{2+}]_i$ have been detected with light-emitting probes. Since these changes are entirely intracellular they would have no electrophysiologic effect on the cell as a whole, unless the cyclic changes in $[Ca^{2+}]_i$ were linked in some way to the transfer of charge across the sarcolemmal membrane (e.g., by Na/Ca exchange). There is no evidence that such experimentally induced oscillations of $[Ca^{2+}]_i$ have any relevance to human arrhythmias associated with ischemia.

POTASSIUM CURRENT IN HYPOXIA/ISCHEMIA

In the earliest stage of hypoxia APD-shortening is not blocked by glibenclamide, and may be produced by an adrenergically activated chloride current. After more prolonged hypoxia/ischemia APD-shortening is prevented by sulfonylureas, even though cytosolic ATP concentration is many times greater than the threshold for activation of ATP-sensitive K^+ channels. Nevertheless, because the sulfonylureas block the hypoxia-induced K^+ current, it is widely assumed that this current must be $I_{K\text{-}ATP}$. The ATP-sensitive channel is

one of a family of ATP-binding channels. "Cystic Fibrosis Transmembrane-conductance Regulator (CFTR) and other ATP-binding cassette-containing proteins may be regulators of channels and pumps" (16). The sulfonylurea receptor (SuR) from pancreatic β cells was cloned and coexpressed with an inwardly rectifying K^+ channel, Kir6.2, to reconstruct an $I_{K\text{-}ATP}$ channel, the conductance of which was blocked by ATP and sulfonylureas, and opened by diazoxides (17). Perhaps hypoxia induces the production of an endogenous K^+-channel opener, which can dissociate the $I_{K\text{-}ATP}$ channel from block by ATP. [ATP-sensitive channels are opened by rapid ventricular pacing in the absence of ischemia (18).]

There are now many drugs classified as K-channel openers, used primarily as vasodilators. The action of pinacidil and cromakalin in dilating larger arteries is endothelium dependent. Aprikalim selectively dilates coronary arteries without affecting blood pressure, independently of endothelium (19). Although in experimental preparations K-channels are opened in cardiac muscle, most of the drugs are 10 to 100 times more selective for blood vessels.

Cibenzoline reduces hypoxic shortening of APD in rabbit hearts, and this action has been observed in rat ventricular myocytes also, and attributed to block of $I_{K\text{-}ATP}$ (20). In patch-clamp studies in guinea pig ventricular cells, flecainide also blocked ATP-sensitive potassium channels (21). In guinea pig isolated perfused hearts the K-channel opener pinacidil reduced the effect of class 1 antiarrhythmic drugs in widening QRS, apparently by reducing the availability of inactivated sodium channels to blockade (22), a result that emphasizes that a drug may have effects in addition to its primary action. A cardioselective K-channel opener, by shortening APD and ERP, could increase the probability of arrhythmias, and would be negatively inotropic by an indirect class 4 action. Hypoxia shortens APD in the atrioventricular (AV) node, and delays AV conduction, again presumably by an indirect class 4 action. Experimentally induced hypoxic AV block is relieved by 50 mM glucose, and by gliben-clamide (23). Whether the direct effect on cardiac muscle of K-channel openers could be beneficial is uncertain. "Potassium channel openers (PCO), which activate cardiac I_{K-ATP}, have demonstrated both antiarrhythmic and proarrhythmic activities in various experimental settings" (24).

ISCHEMIC PRECONDITIONING

In attempts to elucidate the mechanism by which arrhythmias are induced on reperfusion after a period of ischemia, hearts were subjected to ischemia for various lengths of time, followed by intervals of normoxia. It was found that if a brief episode of ischemia, followed by a normoxic interval, preceded a longer period (20–30 min) of ischemia, arrhythmias during and after the long period were reduced. If permanent coronary ligation followed the short period of ischemia, infarct size was reduced. This protective effect of brief prior exposures to ischemia has been termed "preconditioning," and the mechanisms to which the protection might be attributed have been reviewed by Parratt (25), who listed eight main possibilities:

1. Blood flow may be improved (reactive hyperemia).
2. Energy sources may be conserved. The prior depression of contractions conserves ATP, especially if the depression persists ("stunning").
3. Limitation of acidosis.
4. Reduction of oxygen free-radicals.
5. Release of endogenous protective agents [adenosine, nitrous oxide (NO), catecholamines] leading to involvement of phospholipase enzymes, G proteins, proteinkinase C, protein phosphorylation.
6. Reduced release of potentially injurious substances (e.g., noradrenaline).
7. ATP-sensitive channels.
8. Induction of protective enzymes and proteins.

In rabbits subjected to preconditioning (four successive 5-minute occlusions of the anterolateral branch of the circumflex coronary artery interspersed with normoxic periods) infarction induced later by prolonged occlusion was reduced in size from 50% to 31% of the area at risk, but not if the animals were pretreated with an inhibitor (chelerythrine Cl) of PKC. It was concluded that the cytoprotection required activation of PKC (26). In isolated rat hearts, preconditioning (three periods of 3-minute ischemia separated by 5-minute reperfusions) reduced the size of infarcts induced by 30-minute ischemia followed by 60-minute reperfusion. The dependence of the effect on activation of PKC was confirmed, but it was not secondary to stimulation of α-adrenoceptors or of adenosine receptors (27). Preconditioning reduces the fall of intracellular pH during the long ischemic period, and also activates PKC. If PKC is activated by another mechanism, the fall of $[pH]_i$ in ischemia is still reduced, but not by as much as after preconditioning, so that activation of PKC is not the sole basis for the protective effect (28).

Infarct size following 20-minute coronary artery occlusion in rats was reduced from 59% ± 3% to 26% ± 5% by pretreatment with 0.3 mg·kg⁻¹ acetylcholine IV 10 minutes before the occlusion. The effect was NO-dependent, and was blocked by atropine, but not by glibenclamide, which illustrates that I_{K-ATP} and I_{K-Ach} channels are different (29). Other vasodilators also reduced infarct size after ligation of the left anterior descending (LAD) coronary artery in pigs. Nicorandil and nitroglycerin were administered in doses producing a comparable hypotensive effect, and both reduced infarct size, but during the occlusion VT occurred in five of nine animals given nicorandil, but in only three of nine after nitroglycerin (30).

The relative durations of the periods of ischemia and reperfusion are critical in determining the efficacy of the protection provided by preconditioning. In isolated rat hearts a brief (3-minute) episode of ischemia followed by 6-minute reperfusion caused a reactive hyperemia, and an overshoot (115–125%) of the formation of phosphocreatine (PCr). If the second long period of ischemia was started while the PCr was still high, there was a protective effect. If, however, after the initial ischemic episode, the reperfusion was long enough for the PCr overshoot to have declined, the protection was lost (31). Similar findings in anesthetized dogs were reported by O'Connor et al. (32). A brief period of hypoxia followed by 20-minute reoxygenation reduced the number of ectopic beats during a second exposure to hypoxia, but not if the intervening normoxic period was extended to 40 or 60 minutes (32). Brief ischemic episodes led to the formation of heat-stress proteins (HSP) of 70 kd, and it has been suggested that these may be involved in the protective effect of preconditioning (33). Cohen et al. (34) provided evidence that hypoxic preconditioning involved the release of adenosine and noradrenaline, but activation of PKC is not dependent on stimulation of α-adrenergic or adenosine receptors (27).

NONISCHEMIC PRECONDITIONING

The protective preconditioning insult need not be a prior episode of hypoxia or ischemia. Rapid ventricular pacing reduces the incidence of ventricular arrhythmias during a subsequent sustained period of ischemia and reperfusion. Thirty minutes of pacing at 200 beats·min^{-1} immediately preceding coronary occlusion in open-chest pigs reduced infarct size, but not after pretreatment with glibenclamide. The pacing did not produce ischemia; PCr and ATP levels and arterial-coronary sinus differences in pH and pCO_2 were unchanged, and there was no coronary hyperemia at the end of pacing. If 15 minutes elapsed at sinus rhythm at the end of pacing, there was still some residual protective effect, not blocked by glibenclamide, so that opening of I_{K-ATP} channels was not solely responsible for the protection (18).

Nonischemic preconditioning protection against 30 minutes of ischemia plus 120 minutes of reperfusion in rabbit hearts was induced by prior administration of endothe-

lin-1 (ET-1), which activates PKC. Ischemic preconditioning by 5-minute ischemia plus 10-minute reperfusion reduced infarct size from 30.3% ± 2.5% to 5.6% ± 0.7%; 50 μM ET-1 plus nicardipin (to prevent vasoconstriction) reduced it to 5.8% ± 1.0%. Endothelin was not involved in the 5-minute ischemic preconditioning, which persisted in the presence of the ET-1 blocker PD 156707 (35). Even the prior administration of *Escherichia coli* toxin protects against subsequent ischemia plus reperfusion–induced arrhythmias and limits infarct size. The most likely mechanisms appear to be the induction of protective enzymes or proteins, e.g., nitric oxide synthase, cyclic oxygenase (COX)–2 probably mediated by cytokine release (36).

In conclusion a preconditioning protective effect against subsequent ischemia plus reperfusion–induced arrhythmias and infarction can be initiated by a variety of stressful situations, including prior short episodes of ischemia or hypoxia, by rapid pacing, or by exposure to toxins, and may involve activation of PKC, vasodilatation, and increased production of HE phosphates. It is doubtful if such protective stresses could be safely provoked in patients prior to potentially arrhythmogenic procedures such as coronary angioplasty.

PHYSIOLOGIC CARDIAC HYPERTROPHY

Cardiac muscle cells lose the ability to divide soon after birth, and thereafter all growth is by cell enlargement. In rats the transition from hyperplasia to hypertrophy takes place within a period of hours, between the 3rd and 4th postnatal days (37). Adult humans can further enlarge their hearts in response to physical training, mainly by adding sarcomeres, so that cells increase in length more than in width, while retaining orderly conduction and contraction. It is doubtful whether under normal training (unassisted by drugs or hormones) the borderline can be crossed between physiologic and pathologic hypertrophy, to create an "athlete's heart," more susceptible to arrhythmias. Cyclists and other sportsmen have experienced sudden deaths, some attributable to drug abuse, others unexplained. The unexpected incidence of sudden death in orienteers pushed to the limit of their performance was described in a previous chapter. The deaths were attributed to excessive exertion, associated in some cases with a prior infection that had left a residual mild myocarditis. When proper rest periods were introduced in the course of training, no further deaths occurred.

In skeletal muscle the relation between initial length and force of contraction is attributed to overlap between actin and myosin filaments, maximum force being exerted when all myosin heads have actin binding sites in range. In cardiac muscle ventricular output was shown to be proportional to input pressure (preload) long before the ultrastructure of "striped" muscle had been elucidated. Starling employed heart-lung preparations devoid of reflex control by nerves and hormones. The relation between input pressure and stretch is complex, papillary muscles being radially oriented, epicardial fibers circumferentially, and many trabecule at various angles in between. The force of cardiac contraction is mainly controlled by the amount of calcium released from the SR, which is in turn controlled both by the rate of uptake and by the triggering of release by calcium entering the cell through voltage-gated channels. In addition, beat-to-beat control is exercised by autonomic reflexes via arterial and other baroreceptors, and by various messengers of extra- and intracardiac origin. Identification of the signal or signals that initiate cardiac hypertrophy requires examination of many possibilities.

Stretch

In isolated rat ventricular myocytes, stretch caused an immediate increase in contractile force, followed by a slower increase, neither associated with any change in diastolic $[Na^+]_i$ or $[Ca^{2+}]_i$, implying absence of change in Na/Ca exchange. It was concluded that stretch may have increased the accessibility of

myofilaments to Ca^{2+} (38). In ferret ventricle it was concluded from length-force measurements and estimates of $[Ca^{2+}]_i$ by aequorin that the affinity of troponin for Ca^{2+} was length-dependent (39). Reviewing the factors allegedly responsible for stimulating cardiac hypertophy in response to mechanical overload, Mondry and Swynghedauw (40) concluded, "Initially, the effects of circulating hormones, such as catecholamines or angiotensin II, were believed to be the main, and even unique, cause of the modifications in genomic expression. However, there is now evidence that mechanical factors, such as telediastolic volume or wall stress, are the prime movers in triggering cardiac hypertrophy."

In experiments in which right intraventricular pressure was increased by obstruction of the pulmonary artery for 1 hour, Rozich et al. (41) observed no change in plasma catecholamines, yet genes connected to growth were "switched on."

"Aortic banding resulted in a molecular switch in myosin heavy chain expression from α-myosin to the related but distinct β-myosin. Increased afterload of myocytes (stretch) results in the release of angiotensin II which acts both in an autocrine fashion on the stretched cell itself and in adjacent cells via a paracrine mechanism. Cardiomyocytes engage in mitotic activity for only a few days after birth. The mechanism whereby myocytes irreversibly leave the cell cycle is not known" (42).

Catecholamines

In many animal models of cardiac hypertrophy intraventricular pressure has been raised by restricting aortic or pulmonary flow with a ligature (banding). The procedure is acute and traumatic, and the hypertrophy induced is overtly pathologic, with histologic evidence of disruption of myocardial architecture. Large and repeated doses of isoprenaline induce cardiac hypertrophy, causing necrosis and eventual heart failure. After daily intraperitoneal injections of isoprenaline $5 mg \cdot kg^{-1}$ in rats, the heart/body weight ratio increased by 22%, and large changes were induced in APD, which was reduced in the ventricular endocardium from 126 to 91 ms, in the midventricular region from 112 to 98 ms, and increased in the epicardium from 96 to 108 ms (43). The epicardial APD was now longer than the endocardial, reversing the normal relation, and increasing the probability of arrhythmia by reflection.

In other models attempts were made to achieve a more gradual cardiac hypertrophy. In a series of severe experiments rats were forced to run for long periods on treadmills, spurred on by electric shocks if they lagged behind to the bottom of the mill. In another method rats were dropped into water-filled troughs where they had to swim for hours daily for as long as 15 weeks ("training") (44). From the latter experiments, in which cardiac hypertrophy was associated with a high turnover of noradrenaline, and was reduced by injections of guanethidine (chemical sympathectomy), it was concluded that "the results are not really compatible with the work-load hypothesis, but are consistent with the idea that the cardiac sympathetic nerves release a substance(s) that is necessary for the induction of the adaptive hypertrophy of the heart evoked by chronic exercise" (45). This necessary connection led to the view that in all types of hypertrophy cardiac sympathetic nerves were the final common pathway for myocardial cell growth (46).

In another model of cardiac hypertrophy, noninvasive and gradual, rabbits were enclosed in a large transparent box supplied with various mixtures of O_2 and N_2 to simulate ascent to any desired altitude. A purely right-sided ventricular hypertrophy was induced. Electrophysiologic studies revealed that the only significant effect was a uniform and homogeneous prolongation of APD in atrial and ventricular muscle and in Purkinje cells, analogous to a class 3 antiarrhythmic action. Moreover, when the hypertrophied tissues were exposed to acute hypoxia, APD shortened less than in controls. "It was concluded that the hypertrophy did not cause associated electrical alterations likely to increase the risk of arrhythmias" (47).

In young rabbits exposed to hypoxia at 70 to 80 torr O_2 during 9 to 11 days, right ventricular dry weight increased by 57% compared to normoxic controls. The hypertrophy was unaffected by injection of propranolol 5 mg·kg^{-1} twice daily SC, or by 12 mg·kg^{-1} atenolol once daily (48). Pretreatment of the rabbits with guanethidine or 6-hydroxydopamine, at doses that completely abolished responses to sympathetic nerve stimulation, had no effect on the hypoxia-induced hypertrophy (49), nor did treatment with high doses of the calcium antagonists verapamil (5 mg·kg^{-1}) or nifedipine (1 mg·kg^{-1}) twice daily (50). Östman-Smith (51) employed a similar method in rats, and confirmed that prolonged hypoxia induced a right-sided hypertrophy. Rats treated with 80 mg·kg^{-1} propranolol daily exhibited a small, but statistically significant, reduction in the hypertrophy. "The dose of propranolol might seem very large in comparison with doses used in clinical medicine" (51). Propranolol in high doses has a significant Class 1 action, being about four times more potent than procaine as a local anesthetic on frog nerve. No control experiments were undertaken with drugs possessing comparable class 1 action, but without beta-blocking activity, and the small reduction in hypertrophy may be attributed to the negative inotropic effect of 80 mg·kg^{-1} propranolol, rather than to beta-blockade.

In contrast to the homogeneous prolongation of APD by hypoxia-induced hypertrophy, left ventricular hypertrophy following banding of the aorta in cats caused a "greater dispersion of effective refractory period (35 ± 11 versus 12 ± 4 ms, $p < .01$) and of monophasic action potential duration at 90% repolarization (69 ± 25 versus 39 ± 7 ms, $p < .02$)" (52). The animals with hypertrophied hearts were "significantly more vulnerable to [electrically induced] VF, with more spontaneous VF and lower VF thresholds."

The above evidence suggests that cardiac hypertrophy induced by exercise or exposure to high altitude is a response to increased load, and although sympathetic stimulation may be a contributory factor, it is not a necessary one, and mechanical stress alone provides a sufficient signal to switch on protein synthesis. The changes induced by physiologic hypertrophy are not arrhythmogenic, but procedures such as aortic or pulmonary banding cause myocardial disruption and dispersion of APD, which do increase the probability of arrhythmia.

PATHOLOGIC CARDIAC HYPERTROPHY

Some forms of cardiac hypertrophy are secondary to a genetic defect. In the condition known in the United Kingdom as hypertrophic obstructive cardiomyopathy (HOCM) and in the United States as idiopathic hypertrophic subaortic stenosis (IHSS), the hypertrophy was originally attributed to outflow tract obstruction, and was sometimes treated surgically. The condition may, however, exhibit no features of obstruction.

"Outflow tract obstruction, if present, may persist only for a part of a long natural course. Family studies, now sometimes aided by genetic identification of a responsible mutant gene, indicate a wide range of morphologic abnormality not only between individuals and at different ages, but between families. The disorder is heterogeneous—genetically, morphologically, and clinically. Distinguishing it from adaptive hypertrophy may be difficult. Although what is now called myocardial disarray is most rampant in the greatly thickened ventricular septum of typical cases, its extent and location differ widely. What was a rare disorder with a bad prognosis and the cause of tragic sudden death in apparently fit young people has become a common disorder, with a better prognosis in middle age than coronary artery disease" (53).

Sudden death in the condition is most commonly from cardiac arrhythmia, with the implication that it is not the hypertrophy itself that is the main factor, but the associated disarray. Both in primary cardiac hypertrophy, and in hypertrophy secondary to hypertension, the density of β-adrenoceptors is reduced (54).

Genetic defects causing myocardial disease occur not only in nuclear DNA, but also in mitochondrial DNA. All human mitochondrial DNA (mtDNA) is believed to be derived from a single matriarchal "Eve," because of the close homologies in all races. "During evolution from yeast to mammals mtDNA reduced to one-fifth of its size by loss of introns" (55). Programmed cell death, apoptosis, is essential for morphogenesis and development from embryo to adult.

The close relationship $(r = .93)$ between hydroxyl-radical damage and deletion, leading to a progressive decline of the bioenergetic activation of cells and organs, implies an underlying control mechanism of cell death related to the aging process. The major target of the active cell death machinery would be mtDNA. Germ-line point mutations in patients with mitochondrial cardiomyopathy diseases potentially accelerate changes leading to premature aging (55).

Since the capacity for cell division disappears soon after birth, when myocardium is lost (e.g., after MI) cardiac output can only be fully restored by hypertrophy of the surviving cells, but this remodeling may increase disorder of conduction, with greater risk of arrhythmias. Ultimately, the compensation may prove inadequate, and the patient declines into failure. Although positive inotropic agents may afford temporary relief, in the long term they are deleterious, and the only hope for survival is to reduce cardiac work, by lowering heart rate and vascular resistance (56).

ACE INHIBITORS

The Consensus Trial Study Group (57) demonstrated that the addition of the ACE inhibitor enalapril to conventional treatment, including the use of other vasodilators, reduced mortality and improved symptoms in patients with severe congestive heart failure. The efficacy of enalapril was confirmed in later trials (58). These successes suggested that ACE inhibitors might have beneficial effects additional to the lowering of blood pressure. In patients with angina pectoris enalapril 2.5 mg IV had no effect on heart rate, mean blood pressure, and left ventricular end-diastolic pressure, or on AH or HV conduction, but shortened ventricular activity duration (VAD), defined as "the interval from the first point of ventricular activity to the final point of ventricular activity on the ventricular electrogram." VAD was reduced by enalapril from 110 ± 11 to 88 ± 13 ms, implying faster conduction and better cell-to-cell coupling (59). In vitro experiments on rat papillary muscle had revealed that enalapril reduced intercellular gap junction resistance. This effect could be antiarrhythmic by restoring and facilitating conduction if gap junctions had been reduced or disturbed in a remodeled myocardium.

Selective AT_2 receptor blockade does not affect hypertrophy; AT_2 receptors control vascular outgrowth. AT_1 receptors mediate intracardiac renin-angiotensin involvement in normal healing (collagen deposition, interstitial cell DNA synthesis). Selective AT_1 receptor inhibition may restore a proper balance between vascular and muscle growth (60). In spontaneously hypertensive rats the left ventricle hypertrophies, reverting to a fetal-type myocardium. Comparison of the effects of an ACE inhibitor and a selective AT_1 antagonist led to the conclusion that AT_1 receptors were involved in signaling the hypertrophic response to hypertension (61). ACE inhibitors not only decrease the concentration of angiotensin II, but increase the level of bradykinin. In the presence of endocardium, but not in its absence, bradykinin increases tenfold the release of noradrenaline by electrical stimulation, an effect that is augmented by ACE inhibitors (62). The action of ACE inhibitors in modulating sympathetic activity is complex, reducing facilitation by angiotensin II at some sites, and increasing the facilitation by bradykinin at others.

HEART FAILURE

Patients in heart failure often (up to 50%) die of ventricular arrhythmias before the ces-

sation of pump function. This may reflect increasing disarray of conduction, and dispersion of repolarization, in severely damaged ventricular muscle. Specific cellular functions also degenerate. "The positive inotropic and cAMP-elevating effects of both β-adrenoceptor agonists and phosphodiesterase inhibitors are diminished." The messenger RNA level of the inhibitory G protein α subunit, $G_{1\alpha-2}$, is increased in end-stage heart failure, which may account for the reduced response to β-adrenoceptor stimulation (63). Responses to α_1- and β_2-adrenoceptors are also reduced in human ventricular preparations isolated from explanted hearts of patients with end-stage idiopathic dilated cardiomyopathy (64). Inotropic responses to ouabain or raised extracellular calcium were not affected. There was a reduction in the density of β_1-, but not of β_2-adrenoceptors in the failing hearts. It was concluded that, in addition to the downregulation of β_1-adrenoceptors, "impairment of other processes such as the coupling between receptor and effector systems, i.e., the respective G proteins, is equally important in end-stage heart failure" (64). Although responses to adrenergic stimulation are already reduced, depressing them further by beta-blockade may be beneficial. "Experience accumulated from large trials strongly suggested that β-blockers should be used for the management of chronic heart failure. It is appropriate to add β-blockade to conventional therapy, such as diuretics and ACE inhibitors" (65).

In midmyocardial cells isolated from dogs in which heart failure had been induced by rapid ventricular pacing for 3 to 4 weeks, APD_{90} was increased from 842 ± 56 to 1097 ± 73 ms, an effect that was partly attributed to a downregulation of I_{TO} channels (66). In patients with heart failure or with left ventricular hypertrophy, there was a greater dispersion of QT intervals, and increased extrasystoles, in both conditions. VT was infrequently observed, however, in this small study (3 out of 14 with hypertrophy, 1 out of 18 with heart failure) (67).

CONCLUSION

The increased risk of arrhythmias in hypertrophy and heart failure cannot be attributed to any particular abnormalities of cellular function, numerous though they be. The most probable explanation is a gradually increasing disorganization of intercellular conduction, associated with localized cell death. Physiologic cardiac hypertrophy does not appear to be arrhythmogenic. Heterogeneous failure of responses to adrenergic stimuli may increase dispersion of repolarization, facilitating reentry. The beneficial effects of Ace inhibitors are due not only to reduced vascular resistance and cardiac work, but also to a modulation of remodeling in hypertrophy, conserving a favorable balance between cellular and vascular growth.

REFERENCES

1. Kottkamp K, Budde T, Lamp B, Haverkamp W, Borggrefe M, Breidhardt G. Clinical significance and management of ventricular arrhythmias in heart failure. *Eur Heart J* 1994;15(suppl D):155–165.
2. Vaughan Williams EM, Whyte JM. Chemosensitivity of cardiac muscle. *J Physiol* 1967;189:119–137.
3. Gasser RNA, Vaughan-Jones RD. Mechanism of potassium efflux and action potential shortening during ischemia in isolated mammalian cardiac muscle. *J Physiol* 1990;431:713–741.
4. Woodcock EA, Lambert KA, DU X-J. Ins (1,4,5) P_3 during myocardial ischemia and its relationship to the development of arrhythmias. *J Mol Cell Cardiol* 1996; 28:2129–2138.
5. Pucéat M, Vassort G. Neurohormonal modulation of intracellular pH in the heart. *Cardiovasc Res* 1995;29: 178–183.
6. Münch G, Kurz T, Urlbauer T, Seyfarth M, Richardt G. Differential presynaptic modulation of noradrenaline release in human atrial tissue in normoxia and anoxia. *Br J Pharmacol* 1996;118:1855–1861.
7. Petrich ER, Zumino AP, Schanne OF. Early action potential shortening in hypoxic hearts: role of chloride current(s) mediated by catecholamine release. *J Mol Cell Cardiol* 1996;28:279–290.
8. Agetsuma H, Hirai M, Hirayama H, Suzuki A, Takanaka C, Yabe S, Inagaki H, Takatsu F, Hayashi H, Saito H. Transient giant negative T-wave in acute anterior infarction predicts R-wave recovery and preservation of left ventricular function. *Heart* 1996;75:229–234.
9. Remme WJ. Heart failure 95. Multi-author symposium. *Eur Heart J* 1995;16:suppl N.
10. Hemsworth PD, Pallandi RT, Campbell TJ. Cardiac electrophysiological effects of captopril: lack of direct antiarrhythmic effects. *Br J Pharmacol* 1989;98:192–196.
11. Budaj A, Cybulski J, Cedrok K, Karczmarewicz S, Maciejewicz J, Wisniewski M, Ceremuzynski L. Effects of captopril on ventricular arrhythmias in the early and late phase of suspected acute myocardial infarction. *Eur Heart J* 1996;17:1506–1510.
12. Ju Y-K, Saint BA, Gage PW. Hypoxia potentiates persistent sodium current in isolated rat ventricular myocytes. *J Physiol* 1995;489:143P.
13. Greenberg HM, Dwyer EM, Huchman JS, Steinberg JS, Echt DS, Peters RW. Interaction of ischaemia and encainide/flecainide treatment: a proposed mechanism for the increased mortality in CAST. *Br Heart J* 1995; 74:631–635.
14. Ju Y-K, Saint D, Gage PW. Hypoxia increases persistent sodium current in rat ventricular myocytes. *J Physiol* 1996;497:337–348.
15. Diederichs F. Protection of isolated heart against the Ca^{2+} paradox. Are gap junction channels involved? *J Mol Cell Cardiol* 1995;27:1301–1310.
16. Philipson LH. ATP-sensitive K^+ channels: paradigm lost, paradigm regained. *Science* 1995;270:1159.
17. Inagaki N, Gonoi T, Clement JP, Namba N, Inazawa J, Gonzalez G, Aguilar-Bryan L, Seino S, Bryan J. Reconstitution of I_{K-ATP}: an inward rectifier subunit plus the sulfonylurea receptor. *Science* 1995;270:1166–1170.
18. Koning MMG, Gho BCG, van Klaarwater E, Opstal RLJ, Duncker DJ, Verdouw PD. Rapid ventricular pacing produces myocardial protection by nonischemic activation of K^+_{ATP} channels. *Circulation* 1996;93:178–186.
19. Nielson-Kudsk JE, Boesgaard S, Aldershvile J. K^+ channel opening: a new drug principle in cardiovascular medicine. *Heart* 1996;76:109–116.
20. Horie M, Hayashi S, Yuzuki Y, Sasayama S. Comparative studies of ATP sensitive potassium channels in heart and pancreatic β-cells using Vaughan Williams class 1a antiarrhythmics. *Cardiovasc Res* 1992;26:1087–1094.
21. Wang DW, Sato T, Arita M. Voltage dependent inhibition of ATP sensitive potassium channels by flecainide in guinea pig ventricular cells. *Cardiovasc Res* 1995;29: 520–525.
22. Yang Q, Padrini R, Bova S, Piovan D, Magnolfi G. Electrocardiographic interactions between pinacidil, a potassium channel opener, and class 1 antiarrhythmic agents in guinea pig isolated perfused heart. *Br J Pharmacol* 1995;114:1745–1749.
23. Sawardbori T, Adaniya H, Yukisada H, Hiraoka M. Role for ATP-sensitive K^+ channel in the development of A-V block during hypoxia. *J Mol Cell Cardiol* 1995;27: 647–657.
24. Lynch JJ, Sanguinetti MC, Kimura S, Bassett AL. Therapeutic potential of modulating potassium currents in the diseased myocardium. *FASEB J* 1992;6:2952–2960.
25. Parratt JR. Possibilities for the pharmacological exploitation of ischemic preconditioning. *J Mol Cell Cardiol* 1995;27:991–1000.
26. Baxter GF, Goma FM, Yellon DM. Involvement of proteinkinase C in the delayed cytoprotection following sublethal ischaemia in rabbit myocardium. *Br J Pharmacol* 1995;115:222–224.
27. Bugge E, Ytrehus K. Ischemic preconditioning is proteinkinase C dependent, but not through stimulation of α adrenergic or adenosine receptors in the isolated rat heart. *Cardiovasc Res* 1995;29:401–406.
28. Chen W, Wetsel W, Steinbergen C, Murphy E. Effect of ischemic preconditioning and PKC activation on acidi-

fication during ischemia in the rat heart. *J Mol Cell Cardiol* 1996;28:871–880.

29. Richard V, Blanc T, Kaeffer N, Tron C, Thuillez C. Myocardial and coronary endothelial protective effects of acetylcholine after myocardial ischaemia and reperfusion in rats: role of nitric oxide. *Br J Pharmacol* 1995; 115:1532–1538.

30. Klein H, Pich S, Lindert-Heimberg S, Schade-Brittinger C, Maisch B, Nebendahl K. Comparative study on the effects of intracoronary nicorandil and nitroglycerin in ischaemic, reperfused porcine hearts. *Eur Heart J* 1995; 16:603–609.

31. Garnier A, Rossi A, Lavanchy N. Importance of the early alterations of energy metabolism in the induction and the disappearance of ischemic preconditioning in the isolated rat heart. *J Mol Cell Cardiol* 1996;28: 1671–1682.

32. O'Connor PJ, Merrill GF. Ventricular arrhythmias caused by repeated exposure to hypoxia are dependent on duration of reoxygenation. *FASEB J* 1995;9:387–391.

33. Morris SD, Yellon DM, Marber MS. Stress proteins: a factor in cardioprotection? *Heart* 1996;76:97–98.

34. Cohen MC, Walsh RS, Goto M, Downey JM. Hypoxia preconditions rabbit myocardium via adenosine and catecholamine release. *J Mol Cell Cardiol* 1995;27: 1527–1534.

35. Wang P, Gallagher KP, Downey JM, Cohen MV. Pretreatment with endothelin-1 mimics ischemic preconditioning against infarction in isolated rabbit heart. *J Mol Cell Cardiol* 1996;28:579–588.

36. Song W, Furman BL, Parratt JR. Delayed protection against ischaemia-induced ventricular arrhythmias and infarct size limitation by the prior administration of *Escherichia coli* endotoxin. *Br J Pharmacol* 1996;118: 2157–2163.

37. Li F, Wang X, Capasso JM, Gerdes AM. Rapid transition of cardiac myocytes from hyperplasia to hypertrophy during postnatal development. *J Mol Cell Cardiol* 1996;28:1737–1746.

38. Hongo K, White E, Orchard CH. Changes in $[Ca^{2+}]_i$, $[Na^+]_i$ and Ca^{2+} current in isolated rat ventricular myocytes following an increase in cell length. *J Physiol* 1996;491:609–620.

39. Kurihara S, Komukai K. Tension-dependent changes of the intracellular Ca^{2+} transients in ferret ventricular muscle. *J Physiol* 1995;489:617–626.

40. Mondry A, Swynghedauw B. Biological adaptation of the myocardium to chronic mechanical overload. *Eur Heart J* 1995;16(suppl I):64–73.

41. Rozich JD, Barnes MA, Schmid PG, Zile MR, McDermott PJ, Cooper G. Load effects on gene expression during cardiac hypertrophy. *J Mol Cell Cardiol* 1995; 27:485–499.

42. Neyses L, Pelzer T. The biological cascade leading to cardiac hypertrophy. *Eur Heart J* 1995;16(suppl N): 8–11.

43. Shipsey SJ, Ryder KO, Bryant SM, Hart G. Regional action potential changes in isolated ventricular myocytes from rats with catecholamine-induced cardiac hypertrophy. *J Physiol* 1995;489:142–143P.

44. Östman I, Sjöstrand NO, Swedin G. Cardiac noradrenaline turnover and urinary catecholamine excretion in trained and untrained rats during rest and exercise. *Acta Physiol Scand* 1972;86:299–308.

45. Östman-Smith I. Prevention of exercise-induced cardiac

hypertrophy in rats by chemical sympathectomy (guanethidine treatment). *Neuroscience* 1976;1:497–507.

46. Östman-Smith I. Cardiac sympathetic nerves as the final common pathway in the induction of adaptive cardiac hypertrophy. *Clin Sci* 1981;61:265–272.

47. Dukes ID, Vaughan Williams EM. Resistance to hypoxia induced shortening of action potential duration of hypertrophied rabbit hearts. *Cardiovasc Res* 1984; 18:597–603.

48. Dennis PD, Vaughan Williams EM. Hypoxic cardiac hypertrophy is not inhibited by cardioselective or non-selective β-adrenoceptor antagonists. *J Physiol* 1982; 324:365–374.

49. Vaughan Williams EM, Dukes ID. The absence of effect of chemical sympathectomy on ventricular hypertrophy induced by hypoxia in young rabbits. *Cardiovasc Res* 1983;17:379–389.

50. Dukes ID, Vaughan Williams EM. Hypoxia-induced cardiac hypertrophy in rabbits treated with verapamil and nifedipine. *Br J Pharmacol* 1983;80:241–247.

51. Östman-Smith I. Reduction by β-adrenoceptor blockade of hypoxia-induced right-heart hypertrophy in rats. *Br J Pharmacol* 1995;116:2698–2702.

52. Kowey PR, Friehling TD, Sewter J, Wu J, Sokil A, Paul J, Nocella J. Electrophysiologic effects of left ventricular hypertrophy. *Circulation* 1991;83:2067–2075.

53. Oakley CM. Non-surgical ablation of the ventricular septum for the treatment of hypertrophic cardiomyopathy. *Br Heart J* 1995;74:479–480.

54. Chaudhury L, Rosen SD, Lefroy DC, Nihoyannopoulos P, Oakley CM, Camici PG. Myocardial beta adrenoceptor density in primary and secondary ventricular hypertrophy. *Eur Heart J* 1996;17:1703–1709.

55. Ozawa T. Mitochondrial DNA mutations in myocardial disease. *Eur Heart J* 1995;16(suppl O):10–14.

56. Cohn JN and authors of a Veterans Administration Cooperative Study. Effect of vasodilator therapy on mortality in chronic congestive heart failure. *N Engl J Med* 1986;314:1547–1552.

57. The Consensus Trial Study Group. Effects of enalapril on mortality in severe congestive heart failure. *N Engl J Med* 1987;316:1429–1475.

58. Cohn JN. A comparison of enalapril with hydralazine-isosorbide dinitrate in the treatment of congestive heart failure. *N Engl J Med* 1991;325:303–310.

59. González-Fernández RA, Altieri PI, De Mello WC, Escobales N. Electrophysiologic changes in the human heart produced by enalapril. *Am J Ther* 1994;1: 140–143.

60. Smith JFM, Passier RCJJ, Nelissen VR, Ancken HJMG, Cleutjens JPM, Kuizinga MC, Daemen MJAP. Does Ace inhibition limit structural changes in the heart following myocardial infarction? *Eur Heart J* 1995;16(suppl N):41–51.

61. Kim S, Ohta K, Hamaguchi A, Yukimura T, Miura K, Iwao H. Effects of an AT_1 -receptor antagonist, an Ace inhibitor and a calcium channel antagonist on cardiac gene expression in hypertensive rats. *Br J Pharmacol* 1996;118:549–556.

62. Vaz-Da-Silva M, Magina S, Dominguez-Costa A, Moura D, Guimaráes S. The role of the endocardium in the facilitatory effect of bradykinin on electrically-induced release of noradrenaline in rat cardiac ventricle. *Br J Pharmacol* 1996;118:364–368.

63. Eschenhagen T, Mende U, Nose M, Hirt S, Döring V,

Kalmár P, Höppner W, Seitz H-J. Increased messenger RNA level of the inhibitory G-protein α subunit, $G_{i\alpha-2}$, in human end-stage heart failure. *Circ Res* 1992;70: 688–696.

64. Steinfath M, Danielson W, van der Leyen H, Mende U, Meyer W, Neumann J, Nose M, Reich T, Schmitz W, Scholz H, Starbatty J, Stein B, Döring V, Kalamr P, Heverich A. Reduced α_1- and β_2-adrenoceptor-mediated positive inotropic effects in human end-stage heart failure. *Br J Pharmacol* 1992;105:463–469.

65. Cleland JGF, Bristow MR, Erdmann E, Remme WJ, Swedberg K, Waagstein F. Beta-blocking agents in heart failure. Should they be used and how? *Eur Heart J* 1996;17:1629–1659.

66. Kääb S, Nuss B, Chiamvimonvat N, O'Rourke B, Pak PH, Kass DA, Marban E, Tomaselli GF. Ionic mechanism of action potential prolongation in ventricular myocytes from dogs with pacing-induced heart failure. *Circ Res* 1996;78:262–273.

67. Davey PP, Bateman J, Mulligan IP, Forfar C, Barlow C, Hart G. QT interval dispersion in chronic heart failure and left ventricular hypertrophy: relation to autonomic nervous system and Holter tape abnormalities. *Br Heart J* 1994;71:268–273.

9

Proarrhythmia and Combination Therapy

E. M. Vaughan Williams

PROARRHYTHMIA

Although the precipitation of ventricular tachycardia during treatment with quinidine had long been recognized as an idiosyncratic reaction akin to cinchonism, the concept of proarrhythmia, the possibility that antiarrhythmic drugs might exacerbate a preexisting arrhythmia, was presented as an important and identifiable complication of therapy by Velebit et al. (1). Much research was subsequently dedicated to attempts to define the circumstances in which arrhythmias might be aggravated and to identify the drugs most likely to induce proarrhythmic effects (2,3). In a study of 96 patients treated with quinidine or procainamide without any evidence of proarrhythmia, and in whom VT had not been inducible at electrophysiologic study (EPS) prior to drug treatment, 80 remained noninducible, but 16 became inducible.

"Compared with the drug test-negative group, the drug test-positive group had reduced (<40%) left ventricular ejection fractions (80% vs 39%, $p = .005$) and higher prevalence of myocardial infarctions (81% vs 35%, $p = .027$), left ventricular aneurysms (27% vs 5%, $p = .026$) and bundle branch block (53% vs 16%, $p = .005$). Thus, exacerbation of ventricular tachycardia induction after antiarrhythmic agent administration was most common in patients with significant organic heart disease" (4).

In diseased myocardium the presence of scarred nonconducting regions and disruption of the normal pathway of conduction may permit an action potential, further delayed by a Class 1 agent, to arrive at a fully repolarized cell which was previously refractory, and thus to initiate reentry. Some sodium channels recover from inactivation before inward rectification channels open, so that there is a brief vulnerable period (VP) during which an applied stimulus may precipitate a sequence of two or three action potentials, as the first premature response reexcites now fully recovered cells in its neighborhood. In normal guinea pig ventricular myocardium Nesterenko et al. (5) measured the duration of this period as <4 ms. VP was widened to 17 ± 4 ms and to 35 ± 4 ms in the presence of 1 and 12 µM moricizine, respectively, and after 40 minutes in 3 µM flecainide VP was extended to 66 ms. In contrast, quinidine prolonged APD from 128 to 142 ms, and also delayed the most premature boundary (MPB) of activation from 106 ± 10 to 121 ± 12 ms, but shortened the duration of VP to 2 ms (5). The prolonged APD appeared to have eliminated the early part of VP. Class 1c drugs eliminate a proportion of sodium channels, but leave the remainder drug-free, so that they recover from inactivation and can respond to an applied stimulus at the normal time after repolarization. Widening of VP cannot account for the proarrhythmic action of quinidine, but could contribute to that of Class 1c compounds.

In a canine model of myocardial infarction (MI) the effects of flecainide (2 mg·kg⁻¹ infused over 5 minutes plus 1 mg·min⁻¹ thereafter) and of D-sotalol (4 mg·kg⁻¹ plus 2 mg·min⁻¹) were compared in tests of vulnerability to programmed electrical stimulation (PES). Sustained VT was induced in 13 dogs; 12 became noninducible after D-sotalol, but only one after flecainide. In 10 dogs VT could not be induced before drug administration, but all of those in this group that subsequently became inducible had been given flecainide. Signal averaged QRS measurements were digitized and filtered to yield high frequency total duration (HFTD) of the activation time of the ventricle. This was increased from 72.4 ± 16 to 86.5 ± 26 ms by flecainide, but was not significantly altered by D-sotalol (72.5 ± 1.9 vs 74.7 ± 2.4 ms). The 14-ms increase in activation time suggests that the invasion of the ventricle was not only slower after flecainide, but also more widely dispersed. The corrected values only, QT_C and JT_C, not absolute values, were reported, so that the dispersion and widening of QRS cannot be compared with that of QT (6).

Another experimental procedure for measuring the effect of various antiarrhythmic drugs on the "geometry of the activation process" was developed by Dhein et al. (7); 256 unipolar leads, distributed over the surface of isolated Langendorff-perfused rabbit hearts, paced at 150 beats·min^{-1}, mapped the time at which the epicardium was depolarized, that is, the most rapid negative deflection, corresponding to the upstroke of the action potential (8).

"The difference between the activation and repolarization time-points gave the activation-recovery interval (ARI)." The homogeneity of ARI was analysed by determining the standard deviation of ARI in each heart region as "dispersion of ARI." The sequence of activations over the surface provided an "excitation pattern." The first electrode to record a deflection located the "break-through point" (BTP). For each electrode an "activation vector" (VEC) was calculated from the activation times and locations of the surrounding electrodes. These quantitative measurements made it possible to determine how consistent was the excitation from beat to beat under control conditions and after infusion of low, medium, and high therapeutic concentrations of antiarrhythmic drugs.

The activation recovery interval was increased by all the drugs tested: quinidine by 35%; disopyramide, 24%; propafenone, 21.5%; sotalol, 14%; flecainide, 13%; pro-

cainamide, 10%; lidocaine, 9%; mexiletine 0.6%. These results agree with the drugs classification, because ARI could be increased either by a class 1 action, which delays activation, or by a class 3 action, which delays repolarization. Quinidine does both; mexiletine delays activation, but accelerates repolarization, with no net increase in ARI. "The standard deviation of ARI (= ARI dispersion) was increased only by flecainide," from 9.3 ± 0.9 ms (control) by $30\% \pm 11\%$ and $42\% \pm 14\%$ in the presence of the medium and high concentrations respectively. "In contrast, sotalol significantly decreased ARI dispersion" by 13% and 17% at the medium and high concentrations. "ARI dispersion was also decreased by propranolol" by 12.4% at the medium concentration, but the effect was not statistically significant. That the only drugs to reduce dispersion were beta-blockers suggests involvement of endogenously released noradrenaline. The greater dispersion of ARI induced by flecainide was significantly reduced by propranolol administered simultaneously.

The pattern of excitation varied considerably from beat to beat in the control hearts. There were up 15 to 25 breakthrough points (BTP) and in only $72\% \pm 3.5\%$ was the BTP "identical for two heart beats 30 min apart." This variation was increased by all the drugs, the percentages of identical patterns falling (the greater the variation, the smaller the percentage of identical patterns). VEC exhibited even greater variation, with only 25% to 35% identity in controls, and it was increased by all the drugs (7). It would be surprising to find a comparably variable excitation pattern in normal human hearts, which are under beat-to-beat control by autonomic and intrinsic cardiac nerves. The oxygen carrying capacity of the Langendorff perfusion fluid is only a fraction of that of blood, and it lacks the many circulating messengers (adrenaline, angiotensin, insulin, vasopressin, etc.) that influence cardiac function.

"Loss of BTP or VEC similarity indicates alteration of the primary epicardial excitation pattern and is considered to represent an equivalent of proarrhythmic activity." On the assumption that drugs causing greater increases in variability will be most prone to proarrhythmia, the highest arrhythmogenic risk would be assigned to flecainide and propafenone, for different reasons. Propafenone greatly reduced the percentage of identity in BTP and VEC. Flecainide increased ARI dispersion. Although these results cannot be extrapolated directly to a clinical setting, they support other evidence that class 1 drugs may not only delay conduction, but also increase heterogeneity of conduction times, and confirm the distinction between class 1 and class 3 actions in this respect.

In patients who already have disrupted conduction pathways as a consequence of myocardial disease, a class 1 drug might exacerbate heterogeneity to the point of proarrhythmia. Twenty-nine patients "with no prior history of sustained ventricular tachyarrhythmias who suffered a cardiac arrest only while receiving type 1a antiarrhythmic agents (quinidine, procainamide)," were submitted to electrophysiologic study in the absence of the drugs, and 19 had inducible sustained ventricular tachycardia; eight of the patients were noninducible, but were retested after receiving quinidine or procainamide. Only two became inducible. The authors concluded, "Therapeutic management should be as aggressive in this population as that used in patients experiencing sudden death in the absence of drug therapy" (9).

Class 1 drugs increase the threshold for electrical stimulation in isolated tissues, and it has been suggested that concomitant therapy with a class 1 drug might necessitate an increase in the power required to terminate an arrhythmia in patients with an implanted cardiovertor-defibrillator (ICD). A patient with an ICD who was receiving oral procainamide in addition experienced a period of frequent shocks. In a subsequent EPS "procainamide had no effect upon the defibrillation threshold. The patient was discharged receiving disopyramide and mexiletine, and has been free of ICD discharges over 5 months of follow-up" (10). Idiosyncratic responses to particular drugs, including antiarrhythmic drugs, is a factor adding to the unpredictability of proarrhythmic effects.

EXTRACARDIAC INFLUENCES

The extent of intracardiac disruption of conduction by disease is the dominant risk factor for induction of arrhythmias, and modulation by antiarrhythmic drugs, though usually beneficial, might promote an arrhythmia in a particular substrate. Heart function is susceptible to many influences of extracardiac origin, which could also affect the probability of arrhythmia.

Histamine

Anaphylactic reactions may induce cardiac arrhythmias. Atria isolated from guinea pigs previously sensitized with ovalbumin, when challenged with antigen in vitro, exhibit reactions of varying severity, from tachycardia to atrial flutter. Intracellular recording showed a fall in resting potential accompanied by a reduction of V_{max} and prolongation of APD. All these effects could be matched by appropriately elevated concentrations of histamine in the bath fluid (11). The hearts of different mammalian species vary widely in their response to histamine, cats being virtually insensitive, dogs slightly more sensitive. The hearts of guinea pigs and humans respond vigorously. In guinea pigs the histamine contents ($\mu g.g^{-1}$ fresh weight) of the right atrium, right ventricle and left ventricle were 7.12 ± 1.35, 2.3 ± 0.24, and 1.56 ± 0.29, respectively, and in human hearts 1.02 ± 0.07, 0.96 ± 0.19, and 0.71 ± 1.0 (RW Gristwood, 1982, doctoral thesis, London University, unpublished). Both in guinea pig and human hearts the electrophysiologic effects and increased contractions were mediated by H_2 receptors, being blocked by cimetidine.

The effects of H_2-receptor stimulation involved an adenosine $3',5'$-cyclic monophosphate (cAMP)-mediated pathway, but were not identical to those of β_1-adrenoceptor stimulation, which again raises the question of how different receptors achieve separate overall effects, although apparently sharing a common messenger, implying the existence of an intracellular architecture that guides the message from receptor to effector. Many compounds (curare, 48/80) release histamine from mast cells, which are present in the heart. The possibility that histamine might be a contributory factor in proarrhythmia warrants investigation.

Atrial Natriuretic Peptide

Although released from the heart, the primary target of atrial natriuretic peptide (ANP) is extracardiac, in the control of intravascular volume, but it might have a direct cardiac function also. It is released in cardiac failure in quantities that could have diagnostic significance, especially if an extraction process involving the N-terminal is used, providing a more stable product (12). Infusion of ANP into human volunteers had no effect on heart rate or blood pressure, but induced a dose-related reduction in isovolumic relaxation time, by 6 ms at an infusion rate of 5 ng·kg^{-1}·min^{-1}. Such an effect could facilitate ventricular filling in diastole (13). In patch-clamped rabbit heart cells human ANP inhibited L-type calcium current via a mechanism involving guanosine 3′,5′-cyclic monophosphate (cGMP) (14). It is doubtful whether the latter effect would be of sufficient magnitude to constitute an endogenous class 4 antiarrhythmic action, but a facilitation of ventricular filling could improve cardiac function in failure, and so indirectly reduce arrhythmic risk.

COMBINATION THERAPY

Soon after the introduction of sulfonamides it was found that as urine was concentrated in the tubules, filtered drug might reach its limit of solubility and crystallize out. Combination in a lower dosage of two or three drugs with added effects on bacteria solved the problem. The most effective antiarrhythmic drugs exert more than one antiarrhythmic action. Amiodarone, though primarily exerting a Class 3 action, has a noncompetitive Class 2 action, and an effect on sodium channels of the Class 1b type. Two different drugs might have additive antiarrhythmic effects, permitting lower dosage for efficacy, below the threshold for unwanted side effects or proarrhythmic risk.

Class 1 Combinations

In experiments on guinea pig papillary muscle with various class 1 agents Kawamura et al. (15) found that the effects were not always additive. At a constant pacing frequency of 1 Hz the shortening of APD by aprindine was increased by the addition of mexiletine, and the lengthening of APD by disopyramide was reduced by mexiletine. The depression of V_{max} by flecainide was increased by mexiletine, but the effect of aprindine on V_{max} was reduced by mexiletine, which suggests that the rapidly acting class 1b drug can displace aprindine from sodium channels, but not the more firmly bound class 1c compound flecainide. Mexiletine would simply occupy channels still drug-free in the presence of flecainide (15). The effect of disopyramide in guinea pig papillary muscle stimulated at 1 Hz was not increased by mexiletine, but when stimulated at 2 Hz the effect of the two drugs on V_{max} was additive (16).

In patients with ventricular arrhythmias treatment with a combination of quinidine and mexiletine enhanced the efficacy of control of arrhythmias resistant to monotherapy (17,18). In dogs with MI induced by left anterior descending (LAD) coronary ligation, a combination of mexiletine and quinidine had greater antiarrhythmic efficacy than quinidine alone, although the latter increased ERP to the same extent as the combination. The combination prolonged interventricular conduction within the periinfarct zone to an extent greater than did quinidine alone (19).

Class 1 and Class 3 Combinations

In porcine papillary muscle and cardiac Purkinje cells sotalol prolonged APD and ERP proportionately. Mexiletine alone prolonged ERP, but shortened APD. "When the two drugs were combined APD was shortened to the same extent as with mexiletine alone, but refractoriness was further prolonged" (20). In canine cardiac Purkinje cells "combination of

30 μM sotalol with 10 μM mexiletine significantly lengthened premature action potential duration at diastolic intervals <50 ms, while the basic action potential duration at a stimulus frequency of 2 Hz was not affected. The fast time-constant for restitution of the action potential duration was significantly slowed" (21). Thus, the reverse frequency dependence of the lengthening of APD by sotalol combined with the positive frequency dependence of the effect of mexiletine on V_{max} "to provide beneficial electrophysiologic effects expected to provide enhanced antiarrhythmic efficacy and fewer proarrhythmic complications" (21).

Whether the sotalol-mexiletine combination would, in fact, be beneficial in an animal model was tested in conscious dogs in which an infarction had been produced by LAD coronary ligation. Mexiletine alone did not prevent induction of VT by programmed stimulation, but increased the cycle length of monomorphic VT. Sotalol prevented induction of VT in 11/16 dogs, but did not significantly increase the cycle length of the VT in the dogs in which it occurred. The combined drugs both prevented VT induction (again in 11/16) and prolonged the VT cycle length. "The high percentage efficacy of the sotalol-mexiletine combination (88% to 100%), and the fact that the drug-induced slowing of ventricular tachycardia was significantly more marked, indicates that this combination has a beneficial effect" (22).

The efficacy of a Class 1 plus Class 3 combination was tested in 50 patients with sustained VT, by administering low-dose sotalol (205 ± 84 mg·day^{-1}) and quinidine sulphate (1278 ± 1423 mg·day^{-1}) or procainamide (2393 ± 1423 mg·day^{-1}). In 21 out of 46 patients VT was rendered noninducible, and in another 17 inducible VT was modified beneficially, yielding a success rate of 83%. The cycle length of induced VT increased from 324 ± 62 to 432 ± 70 ms. In up to 3 years of follow-up "patients with modified or noninducible tachycardia have a low rate of arrhythmia recurrence," after discharge on the combined therapy (23).

As a sequel to explain the above results, the effects of sotalol plus quinidine were studied with microelectrodes in guinea pig papillary muscle. Quinidine (5 μM), sotalol (6 μM) and Q + S prolonged APD$_{90}$ by 9% ± 1%, 13% ± 1%, and 15% ± 2%, respectively, at a pacing frequency of 3 Hz. ERP was prolonged by Q, S, and Q + S by 18% ± 2%, 11% ± 2%, and 27% ± 2%, respectively, indicating that the combination prolonged ERP relatively more than APD. Sotalol had no significant effect on V_{max}, but reduced the effect of quinidine on V_{max}, which "could be accounted for by sotalol interfering with binding of quinidine to the class 1 binding site," but no evidence was provided that such an interference actually occurred. "The combination of sotalol and quinidine appears to exhibit a unique drug profile that causes a marked increase in ERP with a slight increase in APD accompanied by a slight decrease in V_{max}" (24).

Ipazilide

WIN 54117-4 (ipazilide fumarate) combines class 1 and 3 actions, reducing spontaneous frequency in isolated atria, and depressing V_{max} and prolonging APD. It widens QRS in humans. In various animal models of arrhythmia (aconitine-induced in guinea pigs, post-MI in dogs) ipazilide in doses of 1 to 10 mg·kg^{-1} were more effective as an antiarrhythmic than disopyramide, quinidine, or lidocaine. In patients with congestive heart failure ipazilide induced a dose-related depression of cardiac index, though this was not significant until the plasma concentration reached 0.4 μg·ml^{-1} and above (after doses in excess of 100 mg) (25).

Class 2 and Class 3 Combination

Amiodarone is well known as a primarily Class 3 agent, but was originally introduced as an antianginal drug with an antisympathetic action without being a beta-blocker. A new compound CK-3579 prolongs APD, and blocks β$_1$-adrenoceptors in addition. It exhibited antiarrhythmic activity both in animal models, and in humans under electrophysiologic study (26).

Combined Actions of Existing Class 1 Agents

Reexamination of the effects of some well-known Class 1 drugs by the whole-cell patch-clamp technique in guinea pig ventricular myocytes has revealed actions on other membrane currents, in addition to their effect on sodium channels. Mexiletine, cibenzoline, and quinidine cause some depression of L-type calcium current; procainamide, disopyramide, and flecainide reduce I_{K-ATP}; quinidine, propafenone, cibenzoline, and flecainide restrict I_{TO}. Cibenzoline reduces both I_{K-r} and I_{K-s}, and flecainide restricts I_{K-r} only. Mexiletine has no effect on either component of delayed rectification (27), which may explain, at least in part, why the Class 1b compounds are less effective than quinidine, flecainide, or cibenzoline against atrial arrhythmias. The antiarrhythmic agents have been classified in accordance with their predominant actions, and it has always been recognized that selectivity is limited, and that several drugs exert more than one action. It is difficult, however, to estimate how far results obtained by patch-clamp studies on myocytes can realistically be extrapolated to the human situation.

NEW ANTIARRHYTHMIC DRUGS

Antiarrhythmic drugs continue to be introduced, but there are not, as yet, any new classes of antiarrhythmic action, although there may be new mechanisms for producing the same effect. The compound RP 58866, which selectively blocks I_{K-1}, prolonged QT interval, and exhibited antiarrhythmic activity in isolated rabbit hearts, ostensibly by a class 3 action. However, since the inward rectifier current stabilizes the resting potential during diastole, it is not proven that a class 3 action produced by this mechanism would necessarily provide a reliable antiarrhythmic effect in patients (28). Another compound, GLG-V-13, blocked I_{K-r} and prolonged APD independently of frequency, without any effect on QRS or blood pressure. In a canine model of

arrhythmia (post-MI) it was more effective than lidocaine as an antiarrhythmic. It also reduced heart rate, and delayed AV conduction. A new drug, with a new profile, still exerted a familiar Class 3 antiarrhythmic action, prolonging APD and ERP (29). Another new compound, ICI D7288, reduced heart rate in isolated guinea pig atria, an effect not blocked by atropine, and did not antagonize the positive chronotropic actions of isoprenaline. It was thus a specific bradycardic agent, with a Class 5 action.

AVOIDANCE OF PROARRHYTHMIA BY COMBINATION THERAPY

The classification of antiarrhythmic actions has been based purely on experimental and clinical evidence, exposing the similarities and differences between individual drugs. Although criticized, the classification has survived, and is limited because knowledge is limited (30). A physician who is familiar with their properties and side effects is well placed to make a rational choice from the many available compounds to treat each patient in accordance with the diagnosis. Crijns et al. (31) have shown how a logical use of knowledge of the different classes of antiarrhythmic action can lead to a favorable outcome in the treatment of idiopathic ventricular fibrillation.

The incidence of proarrhythmia is determined primarily by the extent of myocardial disease, though choice of drug is also a factor, class 1c agents carrying the highest risk, class 2 the lowest. In patients with serious organic disease, a strategy for avoiding proarrhythmia would be to set a strict limit on the dosage of any antiarrhythmic drug administered to such subjects. If the arrhythmia was still causing symptoms when the limit was reached, instead of increasing the dose, as was done in CAST (see Chapter 2), another drug could be prescribed, the choice depending on the individual circumstances. Atrial flutter is slowed by flecainide, and may be terminated, but if not, success may be achieved by the addition of sotalol. Supraventricular tachycardias may respond to propafenone alone, but the addition of verapamil, or of a class 5 drug if available, would be a logical supplement. Beta-blockers reduce mortality in post-MI patients by 30%, and if arrhythmias still cause symptoms, low-dose amiodarone should be a safe addition. It is pointless to attempt to suppress ventricular ectopic beats, since their frequency does not consistently correlate with survival (32).

ANCIENT CHINESE REMEDIES

Complex societies existed in China while bison still roamed the plains of North America and Britons painted themselves with woad. Traditional remedies, mostly discovered empirically but also influenced by sympathetic magic, have been subjected to analysis by modern methods, to isolate active principals and elucidate their mode of action. One such compound, obtained from a traditional heart remedy extracted from *Cryptocarya chinensis,* is an alkaloid (–)-caryachine. In Langendorff-perfused rabbit hearts it delayed conduction and prolonged QT interval. In experiments on whole-cell cardiac myocytes concentrations of 0.5, 1.5, and 4.5 μM depressed Na^+ current to 67%, 29%, and 12% of control, respectively. I_{Ca-L} was not affected. It also reduced the 4-AP-sensitive component of I_{TO}, with a small effect on I_{K-1} at the highest concentration. Its antiarrhythmic action is thus a combination of class 1 and class 3 effects (33). Another alkaloid, extracted from *Fissistigma glaucescens,* had very similar properties, reducing V_{max} in rat cardiac cells, being more potent than quinidine. In whole-cell voltage clamp studies it blocked Na^+ current (IC_{50} 0.7 μM), and the 4-AP-sensitive component of I_{TO} was inhibited. There was also a positive inotropic effect, presumably associated with the prolongation of APD, because it was blocked by 4-AP, but not by verapamil, prazosin, or propranolol (34). Both these remedies, never before studied by modern techniques, have class 1 and class 3

antiarrhythmic actions similar to those exhibited by Western antiarrhythmic agents.

Plus ça change, plus c'est la même chose.

REFERENCES

1. Velebit V, Podrid P, Lown B, Cohen BH, Graboys TB. Aggravation and provocation of ventricular arrhythmias by antiarrhythmic drugs. *Circulation* 1982;65:886–894.
2. Leving JH, Morganroth J, Kadish AH. Mechanisms and risk factors for proarrhythmia with type 1A compared with type 1C antiarrhythmic drug therapy. *Circulation* 1989;80:1063–1069.
3. Stanton MS, Prystowsky EN, Fineberg NS, Miles WM, Zipes DP, Heger JJ. Arrhythmogenic effects of antiarrhythmic drugs: a study of 506 patients treated for ventricular tachycardia or fibrillation. *J Am Coll Cardiol* 1989;14:209–215.
4. Sager PT, Perlmutter MPH, Rosenfeld LE, Batsford WP. Antiarrhythmic drug exacerbation of ventricular tachycardia inducibility during electrophysiologic study. *Am Heart J* 1992;123:926–933.
5. Nesterenko VV, Lastra AA, Rosenshtraukh LV, Starmer CF. A proarrhythmic response to sodium channel blockade: modulation of the vulnerable period in guinea pig ventricular myocardium. *J Cardiovasc Pharmacol* 1992;19:810–820.
6. Kidwell GA, Gonzalez MD. Effect of flecainide and D-sotalol on myocardial conduction and refractoriness: relation to antiarrhythmic and proarrhythmic drug effects. *J Cardiovasc Pharmacol* 1993;21:621–631.
7. Dhein S, Müller A, Gerwin R, Klaus W. Comparative study of the proarrhythmic effects of some antiarrhythmic drugs. *Circulation* 1993;87:617–630.
8. Vaughan Williams EM. Relation of extracellular to intracellular potential records from single cardiac muscle fibres. *Nature* 1959;183:1341–1342.
9. Hook BG, Rosenthal ME, Marchlinski FE, Buxton AE, Josephson ME. Results of electrophysiological testing and long-term follow-up in patients sustaining cardiac arrest only while receiving type !A antiarrhythmic agents. *Pace* 1992;15:324–333.
10. Fiksinski I, Martin D, Venditti F. Electrical proarrhythmia with procainamide: a new ICD-drug interaction. *J Cardiovasc Electrophysiol* 1993;5:144–145.
11. Feigen GA, Vaughan Williams EM, Peterson JK, Nielsen CB. Histamine release and intracellular potentials during anaphylaxis in the isolated heart. *Circ Res* 1960;8:713–723.
12. Cleland JGF, Ward S, Dutka D, Habib F, Impallomeni M, Morton IJ. Stability of plasma concentration of N and C terminal atrial natriuretic peptide at room temperature. *Heart* 1996;75:410–413.
13. Clarkson PBM, Wheeldon NM, MacLeod C, Coutie W, MacDonald TM. Acute effects of atrial natriuretic peptide on left ventricular diastolic function. *Eur Heart J* 1995;16:1710–1715.
14. Tohse N, Nakaya H, Takeday Y, Kanno M. Cyclic GMP-mediated inhibition of L-type Ca^{2+} channel activity by human natriuretic peptide in rabbit heart cells. *Br J Pharmacol* 1995;114:1076–1083.
15. Kawamura T, Kodama I, Toyama J, Hayashi H, Saito H, Yamada K. Combined application of class 1 antiar-

rhythmic drugs causes additive, reductive or synergistic sodium channel block in cardiac muscle. *Cardiovasc Res* 1990;24:925–931.

16. Toyama J, Kawamura T, Kodama I. Effects of combined use of class 1 antiarrhythmic agents on V_{max} of guinea-pig ventricular muscles. *Cardiovasc Drugs Ther* 1991; 5:801–804.

17. Duff HJ, Roden DM, Oates JA, Woosley RL. Mexiletine in the treatment of resistant ventricular arrhythmias: enhancement of efficacy and reduction of dose-related side effects by combination with quinidine. *Circulation* 1983;67:1124–1128.

18. Duff HJ, Mitchell LB, Manyart D, Wyse DG. Mexiletine-quinidine combination: electrophysiologic correlation of a favorable antiarrhythmic interaction in man. *J Am Coll Cardiol* 1987;10:1149–1156.

19. Duff HJ, Rahmberg M, Sheldon RS. Role of quinidine in the mexiletine-quinidine interaction: electrophysiologic correlates of enhanced antiarrhythmic efficacy. *J Cardiovasc Pharmacol* 1990;16:685–692.

20. Berman ND, Wang L-Y, Ahmed AA. The cellular electropharmacology of mexiletine combined with sotalol in porcine papillary muscle and Purkinje fibre. *Can J Cardiol* 1990;6:423–430.

21. Varro A, Lathrop DA. Sotalol and mexiletine: combination of rate-dependent electrophysiologic effects. *J Cardiovasc Pharmacol* 1990;16:557–567.

22. Chézalviel F, Weissenburger J, Guhennee C, Jagueux M, Davy J-M, Vernhet L, Simandoux V, Poirier J-M, Cheymol G. Antiarrhythmic effect of sotalol-mexiletine combination on induced ventricular tachycardia in dogs. *J Cardiovasc Pharmacol* 1993;21:212–220.

23. Dorian P, Newman D, Berman N, Hardy J, Mitchell J. Sotalol and type 1A drugs in combination prevent recurrence of sustained ventricular tachycardia. *J Am Coll Cardiol* 1993;22:106–113.

24. Berman ND, Dorian P. Cellular mechanism underlying the efficacy of the sotalol-quinidine combination. *J Cardiovasc Pharmacol* 1993;21:609–614.

25. Gottlieb SS, Cines M, Pressel MD. The hemodynamic actions of the antiarrhythmic agent ipazilide fumarate in patients with congestive heart failure. *Clin Pharmacol Ther* 1994;56:75–81.

26. Argentieri TM, Troy HH, Carroll MS, Doroshuk CM, Sullivan ME. Electrophysiologic activity and antiarrhythmic efficacy of CK-3579, a new class III antiarrhythmic agent with β-adrenergic blocking properties. *J Cardiovasc Pharmacol* 1993;21:647–655.

27. Wang DW, Kiyosue T, Sato T, Arita M. Comparison of the effect of class 1 antiarrhythmic drugs, cibenzoline, mexiletine and flecainide, on the delayed rectifier K^+ current of guinea-pig ventricular myocytes. *J Mol Cell Cardiol* 1996;28:893–903.

28. Rees SA, Curtis MJ. The I_{K-1} blocker, RP 58866, is an effective antifibrillatory agent in isolated rabbit hearts. *Br J Pharmacol* 1992;107:96P.

29. Fazekas T, Carlsson L, Scherlag BJ, Mabo P, Poty H, Palmer M, Patterson E, Berlin KD, Garrison GL, Lazzara R. Electrophysiologic characterization of a novel class III antiarrhythmic agent, GLG-V-13, in the mammalian heart. *J Cardiovasc Pharmacol* 1996;28:182–191.

30. Vaughan Williams EM. Classifying antiarrhythmic actions: by facts or speculation? *J Clin Pharmacol* 1992;32:964–967.

31. Crijns HJGM, Wiesfeld ACP, Posma JL, Lie KL. Favourable outcome in idiopathic ventricular fibrillation with treatment aimed at prevention of high sympathetic tone and suppression of inducible arrhythmias. *Br Heart J* 1995;74:408–412.

32. Kerin NZ, Faitel K. Is the degree of arrhythmia suppression a predictor of survival? *Am J Ther* 1996;3:225–228.

33. Wu M-H, Su M-J, Lee S-S, Lin L-T, Young M-L. Electrophysiological basis for antiarrhythmic efficacy, positive inotropy and low proarrhythmic properties of (-)-caryachine. *Br J Pharmacol* 1995;116:3211–3218.

34. Chang G-T, Wu M-H, Wu Y-C, Su M-J. Electrophysiological mechanisms for antiarrhythmic efficacy and positive inotropy of liriodenine, a natural aporphine alkaloid from *Fissistigma glaucescens*. *Br J Pharmacol* 1996;118:1571–1583.

10

The Clinical Pharmacology of Antiarrhythmic Agents

John C. Somberg

An understanding of the clinical pharmacology of antiarrhythmic agents may aid the physician to effectively utilize these agents in the treatment of patients with cardiac arrhythmias.

ANTIARRHYTHMIC DRUG CLASSIFICATION

The development of an antiarrhythmic drug classification was first proposed nearly 30 years ago (1,2). A variety of classifications have emerged as a result of increasing research with newer antiarrhythmic compounds, as well as the advancement in technology that allows one to specifically record electrophysiologic currents and identify specific ion channels (2,3). Although each of these classifications possess limitations, they each assist in the development of a framework for antiarrhythmics based on the drugs' mechanism of actions. Vaughan Williams proposed an early antiarrhythmic classification based on clinical and electrophysiologic similarities of agents. Today his classification is the most frequently utilized, both in clinical practice and research.

Antiarrhythmic drugs are classified into four major groups, well described in previous chapters. Briefly summarizing the classification, Class 1 antiarrhythmic drugs affect the fast sodium channels by inhibiting the conduction through excitable tissue. This effect is principally through decreasing phase 0 of the action potential. Class 1 antiarrhythmics demonstrate various degrees of effect on con-

duction, refractoriness, and on the action potential–based sodium channel binding kinetics as well as on differing actions on other ion channels. However, the Class 1 agents have a common effect on the Na^+ channel with characteristic kinetics permitting them to be further subdivided into Class 1a, 1b, and 1c. Class 1a agents have a moderate effect on phase 0, depressing slow conduction and prolong repolarization. Class 1b antiarrhythmics are characterized by having a fleeting effect on phase 0 depolarization and conduction while shortening repolarization. Class 1c antiarrhythmics are characterized by producing a marked sustained effect on sodium conductance, prolonging conduction but with no effect on repolarization. The Class 2 antiarrhythmics are β-adrenergic receptor blockers. These drugs depress automaticity and increase the effective refractory period of the atrioventricular (AV) node. This group of antiarrhythmic agents may exert a membrane stabilizing effect at pharmacologic and suprapharmacologic serum concentrations; however, this has little clinical significance. Class 3 antiarrhythmics are characterized as significantly prolonging the action potential. As a result, the repolarization period and refractoriness are prolonged without an appreciable effect on excitability. The Class 4 antiarrhythmics consist of the calcium channel blockers. Their electrophysiologic effects are primarily due to increasing AV nodal refractoriness and slowing conduction in the AV node by blocking Ca^{2+} voltage-dependent channels.

The Vaughan Williams classification has proven to be an effective way of grouping antiarrhythmic drugs. Each antiarrhythmic agent and often its metabolites possess a multitude of electrophysiologic properties that may vary subtly due to additional electrophysiologic effects and differing metabolism and kinetics. Therefore, the response to an individual antiarrhythmic agent of a class may not predict the response to other agents of the class.

ANTIARRHYTHMIC DRUGS

Class 1a Antiarrhythmics

Quinidine

Quinidine is one of the oldest antiarrhythmics available. Quinidine was first developed in the 19th century (4). It continues to be frequently used and has been shown to be effective for both supraventricular and ventricular arrhythmias.

Pharmacology/Electrophysiology

Quinidine is derived from the bark of the cinchona tree. Quinidine is a weak base, and depending on its salt formulation, the percent of quinidine may vary from 60% to 83%. Quinidine is considered to be the prototype of Class 1a antiarrhythmic. Its primary electrophysiologic effect is slowing conduction and increasing the effective refractory period in atrial, ventricular, and Purkinje fibers through sodium channel blockade. Quinidine additionally causes prolongation in action potential (reverse rate dependent) and modifies the rate of rise of phase 0 (5,6). In electrophysiology studies quinidine prolongs the AH interval by influencing the refractory period and AV node conduction. On the ECG, quinidine significantly prolongs the QRS and QT intervals. However, at low doses the AV node effect is variable and its effects on concealed conduction and the vasolytic effects may decrease AV node refractoriness. At higher dose, the predominant effect is AV node conduction delay.

Quinidine possesses α-adrenergic blocking activity. This can result in a reflex tachycardia that may increase the sinus node rate (7). Intravenous administration is generally not recommended other than in acute testing procedures and in emergency situations. Intravenous administration has been used safely (8) when a slow continued infusion technique is employed.

Pharmacokinetics

The bioavailability of quinidine may vary from 43% to 93% depending on the salt formulation. Since quinidine is a weak base the primary area of absorption is in the small intestine. The peak plasma concentration is achieved 2 to 4 hours following administration of quinidine sulfate. The volume of distribution of quinidine is 3 L/kg and follows a two-compartment kinetic model (9). Approximately 70% to 90% of quinidine is bound to glycoprotein (pharmacologically inactive). Quinidine protein binding has been shown to decrease in patients following a myocardial infarction, trauma, or surgery (10–12).

Quinidine undergoes both hepatic and renal elimination. The half-life of quinidine is 6 to 8 hours in patients receiving quinidine gluconate. Several metabolites have been identified following liver metabolism. The principal metabolites are 3-hydroxyquinidine and 2-oxoquinidine. The effects of renal or hepatic dysfunction on the pharmacokinetics of quinidine have not been clearly defined, although the liver appears to play a more important role in quinidine elimination. Marked adjustments in dosage for moderate to severe renal or hepatic disease have not been uniformly required, although higher levels have been noted in cirrhotics and following propranolol administration (13).

The plasma concentration of quinidine is 2 to 4 μg/ml. Patients with concentrations >8 μg/ml may present with clinically significant ECG changes and signs of toxicity.

Clinical Use

Quinidine has been demonstrated to be effective in a variety of atrial and ventricular arrhythmias. Similar to other class I antiarrhythmics, quinidine suppresses PVCs and ventricular arrhythmias in approximately 60% to 80% of patients by Holter monitoring. However, following programmed electrical stimulation (PES) testing, overall efficacy decreases to 30% (14).

Quinidine is effective in treating supraventricular tachycardia, including Wolff-Parkinson-White syndrome, by prolonging the effective refractory period of the accessory pathway in the anterograde direction. Its efficacy in converting atrial fibrillation or atrial tachycardia to normal sinus rhythm is 10% to 20% (15).

Adverse Effects

Quinidine causes mild depression in myocardial contractility and automaticity. Dose-related prolongation of the QRS interval has been demonstrated and AV block can develop with severe quinidine toxicity. Increased ventricular response in atrial fibrillation or flutter may be more likely with quinidine than procainamide because of quinidine's anticholinergic action. A syndrome of quinidine syncope characterized by the development of ventricular tachycardia or fibrillation has been described and estimated to occur in up to 3% of the patients receiving quinidine for atrial fibrillation (16). The mechanism of this syndrome is unknown, but may be related to the development of prolonged Q-T interval, increased QT dispersion, and torsades de pointes ventricular tachycardia (17).

The most common adverse effects following chronic therapy are gastrointestinal, including nausea, diarrhea, abdominal pain, vomiting, and anorexia. Central nervous system toxicity of quinidine (cinchonism) includes tinnitus, hearing loss, visual disturbances, confusion, delirium, and coma.

Allergic reactions may manifest as rashes, fever, immune-mediated thrombocytopenia, hemolytic anemia, and, rarely, anaphylaxis. Quinidine is highly protein bound (about 80%) and may cause bleeding in patients receiving the oral anticoagulant coumadin by displacement of coumadin from its protein binding. Cimetidine has been reported to cause a rise in plasma quinidine concentration (18), and rifampin lowers quinidine concentrations by inducing hepatic enzymes (19). Quinidine has also been shown to cause a significant rise in serum digoxin levels, which may lead to a significant incidence of digitalis toxicity (20). Arrhythmias develop from a differential effect of quinidine on myocardium and Purkinje fibers (21).

Dosage and Administration

Depending on the oral formulation used (sulfate, gluconate, or polygalacturonase) various amounts of quinidine are found in the base formulation. The gluconate and polygalacturonase salts are the slow-release preparations. The immediate-release preparation is usually given every 6 hours and the slow-release preparation is given 8 to 12 hours. Quinidine is orally administered with an initial dose of 600 to 1200 g daily in divided doses. Intravenous administration can be accomplished in a safe and effective manner, but the drug is difficult to administer intravenously (22). The usual dose is 0.25 to 0.31 µg/kg/min of quinidine base. The infusion rate must be slow to avoid hypotension and may need to be terminated due to a symptomatic reduction in arterial blood pressure.

Proarrhythmia

Concern has developed regarding the safety of quinidine use. As with all considerations on the ability of an antiarrhythmic to cause arrhythmia, the clinical context is so critical. The patient with an atrial arrhythmia and no ventricular dysfunction is at considerably less risk than a patient with left ventricular (LV) dysfunction, prior myocardial infarction (MI), or a previous episode of sustained VT. Nonetheless, recent reviews of the literature has suggested an overall increased incidence of adverse outcome on quinidine than off. Coplen and associates (23) report an increased mortality for those patients receiving quinidine, although some of the reported causes of death are hard to associate with a proarrhythmic etiology. Another study found that quinidine has a significant incidence of proarrhythmia, alerting physicians to the possible downside of long-term quinidine therapy. This is further supported by the review of those on antiarrhythmic therapy in the Stroke Prevention in Atrial Fibrillation (SPAF) trial in a retrospective review (24), where type Ia antiarrhythmic therapy increased mortality compared to alternative therapy.

Procainamide

Procainamide was identified in the 1930s as having antiarrhythmic activity. However, it took 20 years before a more stable formulation was introduced.

Pharmacology/Electrophysiology

Procainamide is an analogue of procaine hydrochloride. The cardiac and electrophysiologic activities are similar to other Class 1a antiarrhythmics. However, unlike quinidine, procainamide has no alpha-blocking properties and much weaker vagolytic effects (25).

Procainamide slows conduction along with prolonging the effective refractory period of atrial, AV nodal, and ventricular tissue. Procainamide demonstrates dose-related changes on the surface ECG, which includes a widening of the QRS and at times a widening of the QT intervals (much less pronounced than with quinidine).

Pharmacokinetics

The bioavailability of oral procainamide is 75% to 95% in most patients. Peak plasma concentration following the immediate release preparation occurs 1 to 2 hours post administration. The volume of distribution of procainamide is 2 L/kg. The volume of distribution decreases approximately 25% in patients with congestive heart failure (CHF). In contrast to quinidine, procainamide is minimally protein bound. The elimination half-life of procainamide is 3 to 4 hours, but this may be significantly prolonged (up to 20 hours) in patients with renal dysfunction. The sustained release formulation was developed to increase the half-life (5 to 7 hours) which minimizes the dosing regimen frequency and plasma concentration fluctuations.

Procainamide undergoes metabolism by the liver and is eliminated via the kidneys. Procainamide undergoes first-order pharmacokinetics and acetylation in the liver. The acetylation phenotype, genetically determined, controls the rate of metabolism of procainamide. Approximately 40% of patients are fast and 40% are slow acetylators. The remaining 20% of patients are intermediate. The major metabolite of procainamide following acetylation is *N*-acetylprocainamide (NAPA) (26,27). This metabolite occurs in 16% to 21% of slow acetylators and 24% to 33% of fast acetylators. NAPA is excreted via the kidneys and undergoes minimal metabolism. NAPA has different electrophysiologic properties than the parent compound, behaving like a class III drug, prolonging the action potential duration. The contribution of NAPA to antiarrhythmic activity is small given its low concentrations. However, in patients with renal dysfunction who are rapid acetylators, NAPA accumulates. The recommended therapeutic plasma concentration of procainamide is 4 to 8 μg/ml and of NAPA is 10 to 24 μg/ml. It is incorrect to add the NAPA and procainamide levels together since each has a distinctly different range. Cimetidine has been observed to reduce procainamide elimination.

Clinical Use

Procainamide is an effective agent in the management of both ventricular and supraventricular arrhythmias. Although lidocaine has been preferred for the emergency intravenous therapy of life-threatening ventricular arrhythmias, some patients who fail to respond to lidocaine may be safely and successfully treated with procainamide. Procainamide therapy for patients with the Wolff-Parkinson-White syndrome has produced variable results. Reduction in the number of premature atrial complexes (which often initiate tachycardia) and production of retrograde block in the accessory pathway during atrial fibrillation or flutter are potentially beneficial effects of procainamide. However, procainamide may also have deleterious effects in this syndrome, causing an increase in the antegrade refractory period of the accessory pathway and a lesser effect on the shorter refractory period of the AV node. Rarely, procainamide may facilitate AV nodal reentry and development of paroxysmal tachycardia in patients with the Wolff-Parkinson-White syndrome (28). However, procainamide remains the first-line agent of choice in patients with rapid atrial arrhythmias who may have the preexcitation syndrome.

Adverse Effects

Adverse cardiovascular effects of procainamide include prolonged conduction, negative inotropy, and decreased automaticity. Severe cardiovascular toxicity may manifest as high-grade AV block or intraventricular block with prolongation of the QRS complex (a potential life-threatening toxicity terminating in asystole due to procainamide-induced suppression of automaticity in subsidiary pacemakers). Ventricular tachyarrhythmias and shock may develop from toxic doses of procainamide (29). Excessively rapid administration may produce hypotension that is partially due to a ganglionic blocking action that produces peripheral vasodilation.

Adverse effects following chronic oral use of procainamide include gastrointestinal intolerance (nausea, vomiting, and diarrhea) and skin rash. In addition, up to 40% of patients on long-term procainamide may develop a syndrome resembling systemic lupus erythematosus, with arthralgias, myalgias, fever, pleuropericarditis, and circulating antinuclear antibodies. Virtually all patients eventually develop a positive antinuclear antibody test, but antinuclear antibodies and the clinical syndrome occur sooner among slow acetylators of the drug (27).

Dosage and Administration

In the emergency setting, procainamide is administered intravenously. An initial 5 to 15 mg/kg loading dose is given as a continuous infusion or small boluses (100 mg every 5 minutes to a dose of 1000 mg), followed by a maintenance infusion of 2 to 6 mg/min. It is generally recommended that the intravenous loading infusion should not exceed 25 to 50 mg/min. Throughout the intravenous administration careful monitoring of the blood pressure and ECG must be obtained.

Conversion to oral therapy is based on the total daily dose divided by the appropriate intervals. One half-life (3 to 4 hours on average) should elapse before introducing the oral therapy following cessation of infusion. The introduction of the sustained-release formulation of procainamide has greatly facilitated patient compliance by lengthening the dosing interval to 6 hours. A new formulation of Procanbid can be administered twice daily and still maintain effective drug concentrations while facilitating compliance.

Disopyramide

In 1962 disopyramide was first demonstrated to have antiarrhythmic properties in an animal model and in 1977 disopyramide was approved for ventricular arrhythmias in the United States.

Pharmacology/Electrophysiology

Disopyramide is a synthetic antiarrhythmic that is chemically unlike quinidine or procainamide, although its electrophysiologic properties resemble Class 1a antiarrhythmic agents. The effect on the myocardium and conduction are mediated in part through muscarinic blockade. In animals and humans, the direct depressant effect is offset in part by an increase in heart rate (30). In patients with SA node dysfunction disopyramide may cause significant conduction disturbance (31). In atrial and ventricular muscle both the action potential duration and effective refractory period are increased in a concentration-dependent manner. The electrocardiographic changes observed with disopyramide include widening of the QRS and QT intervals and prolongation of the PR interval.

Pharmacokinetics

Disopyramide is rapidly and almost completely absorbed from the gastrointestinal tract. Peak absorption is achieved within 2 hours using the immediate release formulation. In the liver approximately 25% of the dose of disopyramide is metabolized to a mono-N-dealkyl disopyramide (32). It possesses mild antiarrhythmic activity and may slowly accumulate following chronic use. Disopyramide is principally cleared via the kidneys through glomerular filtration and active tubule secretion. A reduced dose is required in patients with renal dysfunction. The elimination half-life varies from 4 to 10 hours in patients with normal renal function. Protein binding ranges from 5% to 50% for the α_1-acid glycoprotein following nonlinear (zero order) pharmacokinetics (33,34). Therefore, even with a small increase in dose, a significant increase in unbound drug may occur.

The effective therapeutic plasma concentration of disopyramide is 2 to 4 µg/ml. However, in the treatment of ventricular arrhythmias a plasma concentration of 4 to 8 µg/ml may be required (35).

Clinical Use

Similar to quinidine and procainamide, disopyramide is effective in suppressing pre-mature ventricular contractions and ventricular arrhythmias. Controlled trials have demonstrated the efficacy of oral disopyramide in preventing ventricular arrhythmias following an acute myocardial infarction and in preventing recurrent atrial fibrillation after direct-current cardioversion (36,37). Disopyramide has been reported to be effective in the treatment of some patients with refractory ventricular tachycardia.

Since disopyramide slows conduction and prolongs refractoriness in accessory pathways, it has been demonstrated to be effective in the treatment of the Wolff-Parkinson-White syndrome (38). Since the drug markedly decreases left ventricular function due to both a direct as well as an indirect effect, it has found utility in treating patients with idiopathic hypertrophic subaortic stenosis (IHSS) who also manifest arrhythmias. Disopyramide by directly suppressing LV function and increasing afterload and thus decreasing cardiac output has potential benefit in outflow-obstructed patients. However, the electrophysiologic effects in patients with underlying myocardial disease could lead to arrhythmia facilitation and thus disopyramide's use in these patients is problematic. Disopyramide is to be avoided in patients with left ventricular ejection fraction (LVEF) of less than 30%. The incidence of CHF in these patient populations may be as high as 50% (39).

Adverse Effects

The adverse effects of disopyramide are most commonly related to its anticholinergic properties: dry mouth, blurred vision (impaired visual accommodation), constipation, and urinary retention. The latter may be particularly prominent in elderly males. Potential adverse cardiovascular effects are similar to those of quinidine, including AV block, increased ventricular response to atrial flutter or fibrillation, hypotension, and idiosyncratic paroxysmal ventricular tachycardia similar to that reported as quinidine syncope

(40). The negative inotropic potential of disopyramide is considerable and substantially exceeds that of quinidine and procainamide with a significant occurrence of severe cardiac failure in association with its use (39). Precautions should be taken among patients receiving both disopyramide and other negative inotropic agents such as propranolol and verapamil, since the additive cardiac depressant actions may be deleterious.

Dosage and Administration

Disopyramide is available in immediate release and controlled release formulations. Standard dosing includes a loading dose of 300 mg followed by a maintenance dose of 100 to 200 mg every 6 hours. Controlled released preparation allows for a convenient 12-hour dosing interval. Patients with renal insufficiency will require a longer dosing interval.

Class 1b Antiarrhythmics

Lidocaine

Lidocaine was first introduced as a local anesthetic in the 1940s. Further research found antiarrhythmic properties, and now lidocaine is a standard therapy for the treatment of acute ventricular arrhythmias.

Pharmacology/Electrophysiology

Lidocaine is classified as a Class 1b antiarrhythmic. It possesses unique electrophysiologic properties that make it valuable for acute therapy. In isolated tissue lidocaine exerts its major electrophysiologic effect by depressing the action potential amplitude. Purkinje fiber action potential amplitude and conduction velocity are reduced and membrane responsiveness is suppressed (41). Lidocaine has minimal or no effect on the sinoatrial (SA) node or AV node conduction velocity or refractory period (42). In general, lidocaine does not significantly demonstrate changes in the surface ECG, although at times the PR and QRS duration may decrease. Lidocaine shortens the QT interval in some patients by shortening repolarization.

Pharmacokinetics

Due to lidocaine's poor absorption properties and extensive first-pass hepatic metabolism, intravenous administration is the preferred route of administration. Lidocaine undergoes liver metabolism to form two principal metabolites. Each of these metabolites has antiarrhythmic properties and are found in significant concentrations. The first metabolite is monoethylglycine xylidide, which is formed by *N*-deethylation of lidocaine. It has a half-life of 120 minutes and is eliminated by the plasma. Following further metabolism, the second metabolite glycine xylidide is formed. It has a half-life of 10 hours and is excreted principally by the kidneys (90%).

Lidocaine follows a two-compartment model and has a beta half-life of approximately 100 minutes. Lidocaine equilibrates into various compartments of the body having a high blood flow (e.g., liver, kidneys, brain, heart). Conditions such as liver failure, CHF, and renal failure have been shown to alter the disposition of lidocaine and thus require dosage adjustment. The therapeutic plasma concentration of lidocaine is between 2 and 5 µg/ml. Patients with plasma concentrations >6 µg/ml may manifest signs of toxicity. Cimetidine causes a significant rise in serum lidocaine concentration (43).

Clinical Use

Lidocaine is uniquely useful in ischemic situations where it may convert a unidirectional block into a bidirectional block, thus preventing a reentrant ventricular tachycardia. Lidocaine possesses a greater depressant effect on V_{max} in ischemic tissue and it decreases dispersion of refractoriness and the propensity to sustain a ventricular tachycardia.

The utility of lidocaine is that one can achieve therapeutic concentrations rapidly (IV or intramuscularly) to control ventricular arrhythmias, especially in the ischemic milieux. In patients resuscitated from out-of-hospital ventricular tachycardia or fibrillation, lidocaine is as effective as bretylium in preventing recurrent episodes of ventricular arrhythmias (44). In patients with an MI within 6 hours, lidocaine prophylaxis reduced episodes of ventricular fibrillation compared to the performance of placebo (45). Other investigators have not reported advantages of prophylactic lidocaine (46) and some have argued for prophylactic lidocaine in only selected patients (47). A recent review by Yusuf et al. (48) has found an adverse effect of lidocaine when data from multiple trials are pooled. The effectiveness of lidocaine has been questioned as first-line therapy for VT following the report of Pacifico and associates (49), who found lidocaine was effective in terminating only 8% of sustained ventricular tachycardia.

Lidocaine has little effect on atrial tissue and has minimal myocardial depressant properties. Thus, lidocaine can be used more easily than Class 1a or 1c antiarrhythmics in patients with sinus node disease or AV conduction disturbance. However, in patients with severely compromised conduction system lidocaine must be used in conjunction with close electrocardiographic monitoring.

The use of lidocaine in the treatment of atrial flutter entails the complication that lidocaine may lead to an acceleration of the ventricular response. In the Wolff-Parkinson-White preexcitation syndrome, the response to lidocaine may be unpredictable, and in patients with a known short effective refractory period of the accessory pathway, lidocaine may accelerate the ventricular response during atrial fibrillation (50). In a macro reentry tachycardia involving the bypass tract in an antegrade fashion (atrium to ventricle) and then up the His-Purkinje system and AV node to the atrium (retrograde), lidocaine may terminate the arrhythmia in the retrograde component of the circuit.

Adverse Effects

The most common adverse effects observed with lidocaine are those associated with the central nervous system. These include dizziness, drowsiness, tinnitus, paresthesias, and visual disturbance. These effects can be more often observed in geriatric patients and are concentration dependent. Reversal of these effects is often seen following a decrease in the dosing regimen. More severe CNS effects can be seen at higher concentrations and include seizures, confusion, stupor, or coma. A mild peripheral vasodilation may result, though rarely causing hypotension. AV block or sinus node dysfunction is also rare.

Dosage and Administration

Lidocaine is administered intravenously by bolus followed by a maintenance infusion. The therapeutic plasma concentration can be usually achieved by a 1- to 2-mg/kg loading dose (administered as a 50- or 100-mg bolus) followed by a 1- to 4-mg/min continuous infusion. If the arrhythmia is not suppressed, a second bolus of 1 to 2 mg/kg may be administered and the continuous infusion should also be increased to 4 mg/min. Patients who are small (under 50 kg) or who are hemodynamically compromised require a lower bolus and maintenance infusion. Monitoring of plasma concentrations may be helpful in avoiding toxicity with prolonged administration. However, even with caution the higher the administered dose, the greater the risk of developing lidocaine toxicity.

Mexilitene

Mexilitene is an oral analogue of lidocaine. It was approved by the Food and Drug Administration (FDA) in 1986 for sale in the United States.

Pharmacology/Electrophysiology

Mexilitene is a Class 1b antiarrhythmic. The cardiac electrophysiologic effects are similar to lidocaine. Mexilitene depresses the rate of rise of the action potential without prolonging action potential duration. Mexilitene has no effect on the atrial refractory period, sinus node function, or AV node refractory period (51,52). In the His-Purkinje system, both the effective and relative refractory periods are prolonged. Similar to lidocaine, the electrocardiogram is often unchanged.

Pharmacokinetics

Mexilitene is rapidly absorbed following oral administration, and peak absorption is achieved in 4 hours. Mexilitene has a markedly reduced first-pass metabolism. The mean elimination half-life of mexilitene following chronic therapy ranges between 11 and 15 hours in patients with coronary artery disease (53). Mexilitene is primarily eliminated by hepatic metabolism. Approximately 85% of mexilitene is metabolized to inactive metabolites: parahydroxy mexilitene and hydroxymethyl mexilitene. The remainder is excreted via the kidneys. Mexilitene does not significantly accumulate in patients with renal insufficiency (54). Alkalization of the urine results in an increase in the plasma concentration increasing reabsorption in the distal tubule. The normal therapeutic plasma concentration is relatively narrow, 0.7 to 1.6 μg/ml; thus, monitoring of the serum concentration may be useful to prevent adverse side effects.

Clinical Use

Mexilitene is indicated in the treatment of ventricular arrhythmias. It has been shown to be effective in suppressing of PVCs following chronic dosing (55) and drug-resistant ventricular tachycardia (56). The effectiveness of mexilitene as monotherapy evaluated by PES techniques shows efficacy comparable to lidocaine, about 10% on average in the electrophysiologic laboratory (57). Several recent reports have augmented this limited efficacy of the drug by combining mexilitene with quinidine or other Class 1 antiarrhythmic agents (58).

Adverse Effects

The most frequent adverse effects of mexilitene involve the central nervous system and gastrointestinal tract. These are often dose related. They include tremors, nystagmus, blurred vision, dizziness, drowsiness, confusion, ataxia, paresthesia, dysarthria, insomnia, tinnitus, nausea, vomiting, and convulsions. Often the gastrointestinal disturbance can be decreased when mexilitene is taken with food.

Dosage and Administration

Mexilitene is usually administered at 8-hour intervals at dosages of between 200 and 400 mg. Patients begin at 200 mg administered every 8 hours, and the dosage is increased every 3 to 5 days until very early toxic signs develop or until the arrhythmia is controlled.

Tocainide

Tocainide is the second oral Class 1b antiarrhythmic to become available. It was approved by the FDA for ventricular arrhythmias.

Pharmacology/Electrophysiology

Tocainide is a primary amine analogue of lidocaine. The electrophysiologic properties of tocainide are similar to both lidocaine and mexilitene. Tocainide shortens the action potential and effective refractory period in the atrial, AV node, and ventricle, and has minimal effect on the Purkinje fibers (59). Due to its minimal effect on SA and AV nodal conductions tocainide has been reported safe in patients with abnormal AV node conduction (60). Tocainide does not significantly alter ECG intervals.

Pharmacokinetics

Tocainide is nearly 100% bioavailable following oral administration. Tocainide undergoes little presystemic elimination; therefore, peak plasma concentration is achieved within 4 hours. Although food has been shown to delay the rate of absorption, the extent of absorption is unchanged. The mean elimination half-life of tocainide is 11 hours (61). Higher elimination half-lives have been observed in patients with CHF and renal dysfunction, ranging from 13 to 22 hours. Tocainide undergoes partial hepatic metabolism to lactylxylidide and tocainide carbamoyl glucuronide. Each of these metabolites is inactive and is excreted renally, and 35% of tocainide is excreted unchanged. Dosage adjustments are usually required in patients with renal dysfunction. The recommended effective plasma concentration is 3 to 7 μg/ml.

Clinical Use

Tocainide is effective in the suppression of ventricular ectopy after a single oral dose and after chronic oral dosing of 400 to 600 mg every 12 hours (62). A number of studies have suggested that in patients with drug-resistant arrhythmias, efficacy of tocainide may be predicted by response to lidocaine (63). In patients with recurrent ventricular tachycardia it was shown that 11 of 15 patients favorably responded to tocainide therapy (64), though the population studied was small and highly selective. It is possible that tocainide, by shortening the refractory period (65), like lidocaine, may prevent a reentrant tachycardia from being induced by PES techniques. Since tocainide is similar to lidocaine in pharmacodynamic properties, it is not surprising that the drug has been studied extensively in the

postmyocardial infarction period in which it has been effective in reducing ventricular ectopy (66,67). It is also effective in prophylaxis against postinfarction arrhythmias (68), as well as in a postsurgical population (69).

Adverse Effects

Tocainide causes side effects, resulting in 10% to 20% of patients discontinuing therapy (70). The side effects are largely neurologic and include tremors, hot and cold flashes, nausea, dizziness, anxiety, vertigo, diplopia, emesis, paresthesia, and coma. Rash, constipation, and fever have also been reported. Most side effects can be reduced by decreasing the drug dose, but this may reduce antiarrhythmic efficacy. A small but significant incidence of neutropenia and agranulocytosis has been reported. Because of this the drug has been recommended for patients unresponsive to other therapy who possess life-threatening arrhythmias and demonstrable lidocaine-responsive arrhythmias.

Dosage and Administration

Tocainide is available in 400- to 600-mg tablets. The recommended initial dose is 1200 mg divided in three equal doses.

Class 1c Antiarrhythmics

Encainide

Encainide is a Class 1c antiarrhythmic. It was approved by the FDA for the treatment of life-threatening ventricular arrhythmias in 1987 and withdrawn by the company in 1990 except for those patients with a life-threatening event who were responsive only to encainide therapy.

Pharmacology/Electrophysiology

Encainide is a benzanilide derivative possessing local anesthetic activity. Its chemical structure is similar to procainamide; however, its electrophysiologic properties are different. It prolongs conduction markedly in atrial, AV node, and myocardial tissue (71). The drug has less negative inotropic action than other Class 1c agents. The ECG effects observed with encainide include prolongation of the PR and QRS intervals.

Pharmacokinetics

Encainide undergoes two distinct metabolic pathways that are dependent on the individual patient's genetic phenotype. Two patient populations have been identified, extensive metabolizers and poor metabolizers, each possessing different pharmacokinetic profiles. Approximately 93% of patients are extensive metabolizers. The two major metabolites, which are formed following this pathway, are O-dimethyl encainide (ODE) and 3-methoxy-O-dimethyl encainide (MODE), which have greater potency and contribute to the therapeutic, as well as the adverse effects, of encainide (72). Peak plasma concentrations following oral administration occur at 1.5 hours for encainide, 1.4 hours for ODE, and 5.7 hours for MODE. The systemic bioavailability is variable, ranging from 14% to 38% due to the significant first-pass effects. The volume of distribution is 27 L and protein binding at steady state is 70%. The elimination half-life of encainide and ODE range from 30 minutes to 4 hours and 5 to 37 hours, respectively. Both ODE and MODE accumulate significantly following encainide metabolism and are renally eliminated, accounting for 40% of the urinary excretion products. Encainide clearance is decreased in extensive metabolizers with renal dysfunction and both ODE and MODE concentrations may increase (80% and 15%, respectively) following long-term dosing. Therefore, dosage adjustment may be required.

Seven percent of patients receiving encainide are poor metabolizers and form minimal amounts of ODE and MODE. In this patient population encainide is primarily metabolized to N-desmethyl encainide, which has equal antiarrhythmic potency as encainide and is observed in both plasma and urine. The elimination half-life for encainide is much longer, ranging from 8 to 22 hours. In contrast to extensive metabolizers, there is an increase in the bioavailability (approximately 85%) and protein binding (78%).

Due to the complex metabolism of encainide and the activity of the parent compound and metabolites, monitoring of the plasma concentration of both encainide and its metabolites offers minimal advantage. The drug's excessive prolongation of the QRS is a possible means of assessing increased or possibly excessive drug concentrations.

Clinical Use

Encainide is effective in the treatment of supraventricular and ventricular arrhythmias (73). Studies have demonstrated nearly complete suppression of ventricular ectopy in patients with significant arrhythmias (74). The occurrence of proarrhythmia with encainide varies from 2.8% to 16.0% depending on the severity of the arrhythmia and concurrent risk factors. Results from the CAST study (75) find an increase in sudden death with encainide and flecainide compared to patients treated with a placebo postmyocardial infarction when their VPCs were suppressed on Holter monitoring. Encainide should not be used in post-MI patients and in only those patients with a life-threatening arrhythmia who have undergone electrophysiologic testing.

Adverse Effects

The most common side effects are dizziness (26%) and visual disturbances (19%) (76). Other common adverse effects include nausea (11%), headaches (15%), taste disturbances (4%), and tremors (3%). These side effects are responsible for discontinuation of therapy in approximately 5% to 10% of patients.

Dosage and Administration

The initial dose of encainide in patients with normal renal function is 25 mg administered every 8 hours. The dosage may then be subsequently increased to 35 mg every 8 hours and then 50 mg every 8 hours with 3 days or more between increments. This time period is needed since the enzyme system metabolizing encainide to active metabolites take days to be fully expressed. Due to the increased risk of proarrhythmic events at higher doses, it is generally recommended that the lowest dose that suppresses the arrhythmia should be used.

Flecainide

Flecainide was the first Class 1c antiarrhythmic available in the United States. It was synthesized in 1972 and approved in 1986. The drug is indicated for the treatment of life-threatening ventricular arrhythmias as well as the treatment of supraventricular arrhythmias.

Pharmacology/Electrophysiology

Flecainide is a fluorinated aromatic-hydrocarbon. It is distinctly different electrophysiologically from quinidine and procainamide. The drug has prolonged binding kinetics to the sodium channel, decreases the rate of rise of the action potential, and markedly prolongs conduction in the atrial, AV nodal, and ventricular myocardial tissues (77). The electrophysiographic effects of flecainide include significant increase PR and QRS duration. The repolarization component of the QT interval remains unchanged.

Pharmacokinetics

The pharmacokinetics of flecainide have been evaluated following oral and intravenous administration. The oral bioavailability of flecainide is approximately 95%. Peak serum concentrations are rapidly achieved within 90 minutes. Approximately 40% of flecainide is protein bound. Flecainide undergoes hepatic metabolism to form meta-0-dealkylated flecainide and the meta-0-dealkylated lactam of flecainide (78). Each of these metabolites is found in the plasma and excreted in the urine. They possess no clinically significant electrophysiologic or antiarrhythmic properties at normal therapeutic doses.

Following chronic dosing, flecainides half-life ranges from 9 to 23 hours. Approximately 25% of flecainide is principally cleared via the kidneys unchanged. Renal dysfunction and heart failure have been reported to prolong flecainide's half-life (78,79). A correlation has been reported between the plasma concentration of flecainide and suppression of premature ventricular contractions. The recommended plasma concentrations are between 200 and 1000 ng/ml, while plasma concentration >1000 ng/ml may be associated with toxicity.

Clinical Use

Flecainide has been shown to be effective in both atrial and ventricular arrhythmias (80,81). Patients undergoing electrophysiologic evaluation for ventricular arrhythmias have reported a greater overall efficacy with flecainide than with either Class 1a or 1b antiarrhythmics (82). Flecainide is approved for the treatment of atrial arrhythmias, and has been found effective in suppressing ectopic and reentrant atrial arrhythmias, AV nodal reentrant atrial arrhythmias, and AV reciprocating tachycardia (83,84). It should be used for patients without structural heart disease since patients with structural heart disease may be subject to the proarrhythmic effects of the drug.

Flecainide is perhaps the best studied agent in the treatment of supraventricular tachycardia (SVT). Administration of flecainide is reported to increase the number of patients free of arrhythmic attacks of paroxysmal SVT (PSVT) from 24% to 82% and patients with AF from 12% to 68% (85). In addition, the interval between attacks is increased from a median of 12 days on placebo to more than 55 days on flecainide (86). This efficacy appears to be dose related (87). The arrhythmia control is also associated with a significant reduction in symptoms (88).

Adverse Effects

Flecainide possesses negative inotropic activity and it lengthens ventricular refractoriness. Thus, patients with significant congestive heart failure and severe conduction disturbances have relative contraindications to flecainide administration. Patients with a history of sick sinus syndrome often require implantation of a permanent pacemaker (82).

Of considerable concern with the use of flecainide is the potential proarrhythmic effects of the drug. Increased potency seems to also involve an increased proarrhythmic potential. Patients with low ejection fraction and a history of sustained ventricular tachycardia or cardiac arrest seem especially prone to the proarrhythmic effects of the drug. The proarrhythmic effects of the drug may be manifested by the development of a wide complex ventricular tachycardia that may be unresponsive to electroshock, accelerate, become incessant, and lead to death. Patients unresponsive to electroshock and suffering from flecainide toxicity may respond to lidocaine boluses followed by infusion (89).

The recent report of the Cardiac Arrhythmia Suppression Trial showed that flecainide can be well tolerated and suppress VPCs in post-MI patients. However, patients receiving flecainide are at a greater risk of dying than those taking placebo (75). One can hypothesize that the proarrhythmia potential is greater than the arrhythmia risk for this population and the benefit offered by flecainide. A number of commentaries have suggested that the proarrhythmic role of flecainide is due to the interaction of the drug with active ischemia. Ranger and associates (90) have reported the acute widening of the QRS on exercise testing in patients who become ischemic. This may be a way to identify patients at risk for development of wide complex VT.

Other adverse reactions of flecainide include CNS and GI toxicity with headache, drowsiness, dry mouth, nausea, and vomiting reported.

Dosage and Administration

The dosage of flecainide is initiated at 100 mg twice daily. After the patient is on 100 mg twice daily and is loaded over 3 or 4 days, obtaining an appropriate trough serum level can be helpful in guiding therapy. In patients with cardiac dysfunction, a starting dose of 50 mg twice daily is recommended, although the possibility of proarrhythmia in this population may not be dose related.

Propafenone

Propafenone is an antiarrhythmic studied extensively in Europe. It is a Class 1c antiarrhythmic. At doses often higher than those used clinically propafenone demonstrates mild β-adrenergic and calcium channel blocking properties. The drug is indicated only for the treatment of life-threatening ventricular arrhythmias in the United States.

Pharmacology/Electrophysiology

Propafenone has a variety of antiarrhythmic characteristics that are concentration dependent. The principal electrophysiologic activity is depression of the rate of rise of the action potential in the atrium, ventricles, and Purkinje fibers (91,92). Propafenone has no effect on the action potential duration and a minor effect on the effective refractory period. The drug markedly prolongs conduction. Propafenone also decreases automaticity in the SA node, atrium, and Purkinje fibers. In patients with AV node accessory pathways, propafenone prolongs the conduction in both the antegrade and retrograde direction, making it effective in treating pre-excitation-mediated arrhythmias (93).

Following chronic oral therapy in patients with left ventricular dysfunction, propafenone has been reported to further impair LV function, with reductions up to 20% (94). The ECG effects of propafenone are dose dependent. Propafenone prolongs PR and QRS intervals, but does not affect the repolarization component of the QT, JT, and QTc intervals. Placement of intracardiac catheters during electrophysiology studies have reported prolongation of the AH and HV intervals and refractory period of the right atrium, ventricle, and AV node.

Pharmacokinetics

Propafenone is completely absorbed following oral administration. Peak plasma concentration is achieved within 2 to 3 hours. Due to first-pass metabolism, bioavailability is low; however, bioavailability increases with high doses. The volume of distribution of propafenone is 1.9 to 3.6 L/kg and is 95% protein bound. In the liver, propafenone undergoes oxidative metabolism to form metabolic products of glucuronide and sulfate conjugates. Elimination is primarily observed in the feces and urine. The principal metabolites are 5-hydroxypropafenone and N-depropylpropafenone (95). The 5-hydroxypropafenone has been reported to have significant electrophysiologic activity. Slow and fast metabolizers have been identified; this is genetically determined. Approximately 90% of patients receiving propafenone are fast metabolizers and 10% of patients are slow metabolizers (96). The mean elimination half-life for fast metabolizers is 5.5 hours and for slow metabolizers is 17.2 hours.

Clinical Use

Propafenone has been shown to be effective in the treatment of supraventricular and ven-tricular arrhythmias. In supraventricular arrhythmias propafenone has been reported to be effective in AV nodal reentry, Wolff-Parkinson-White syndrome (97), and atrial flutter (AF) and fibrillation (98). Propafenone does not significantly affect the energy requirement to terminate AF while significantly decreasing the rate of recurrence (99). The drug is currently not approved for the treatment of supraventricular arrhythmias in the United States.

In the treatment of ventricular arrhythmias propafenone is effective in suppressing ventricular ectopy and nonsustained VT. In suppressing chronic stable ventricular arrhythmias propafenone has been reported to be more effective and better tolerated than disopyramide, although it has similar efficacy to quinidine. In the treatment of more severe arrhythmias propafenone has been reported to be less effective. Studies in patients with inducible sustained ventricular tachycardia by programmed electrical stimulation propafenone is less effective and has a high proarrhythmic incidence (100). In the Cardiac Arrhythmic Study (Hamburg), the propafenone-treated arm showed a higher mortality than the other groups, suggesting a similar outcome to the CAST results (101).

Adverse Effects

Propafenone possesses myocardial depressant properties similar to flecainide. Propafenone may precipitate or exacerbate congestive heart failure, conduction disturbances, and aggravation of arrhythmias. Additional adverse effects include bitter or metallic taste, constipation, nausea, dizziness, diplopia, paresthesia fatigue, and headache. These effects may occur in as many as 20% of patients, although the incidence can be reduced by decreasing dose. Propafenone, being a Class 1c antiarrhythmic, probably has the same cautions as flecainide, given the CAST results with the Class 1c agents and the results of the Cardiac Arrhythmia Study (101).

Dosage and Administration

Propafenone is available in 150- to 300-mg tablets. The initial dose is 150 every 8 hours. The maximum recommended chronic dosing is 900 mg/day in divided doses.

Moricizine

Moricizine is a Class 1c antiarrhythmic. It was initially developed in the Soviet Union in 1964. By 1976 clinical studies were initiated in the United States that led to its approval by the FDA in 1990 for the treatment of life-threatening ventricular arrhythmias.

Pharmacology/Electrophysiology

Moricizine is a phenothiazine derivative. Its principal electrophysiologic effects are caused by a concentration-dependent blockade of the sodium channels and electrophysiologically acts like a Class 1c antiarrhythmic (102). Moricizine decreases the maximum upstroke velocity of the action potential along with increasing the rate of membrane repolarization. In addition, both the action potential duration and effective refractory period are decreased. In electrophysiologic testing, moricizine markedly slows intraatrial, AV nodal, and intraventricular conduction (103). The electrocardiographic changes observed include prolongation of the PR and QRS intervals. Electrophysiologic study results place the agent in the Class 1c category, though a weaker agent than either flecainide or propafenone (102).

Pharmacokinetics

The pharmacokinetics of moricizine has been described with the oral formulation. Following oral administration, moricizine undergoes rapid absorption and first-pass metabolism (104). Peak plasma concentration is achieved within 1 to 2 hours following administration. Approximately 38% of the total moricizine dose is absorbed. In healthy volunteers the volume of distribution ranges from 8.3 to 11.1 L/kg. Moricizine is significantly bound to 2-glycoprotein, albumin, and peripheral tissue in the range of 95%.

In the liver, moricizine undergoes significant metabolism (105). In animals 30 to 40 metabolites have been reported and in humans a minimum of nine metabolites have been identified in urine and feces. Two metabolites have been shown to possess significant antiarrhythmic activity; these include moricizine sulfoxide and phenothiazine 2 carbon acid ethyl ester sulfoxide. Moricizine and its metabolites are cleared, 56% in the feces and 39% in the urine (106). The plasma clearance ranges from 2.2 to 2.6 L/min in healthy volunteers. The mean elimination half-life of moricizine following chronic dosing is 19.2 hours. Decreases in hepatic blood flow and hepatic dysfunction alter the pharmacokinetics of moricizine, while congestive heart failure does not.

Due to the complex hepatic metabolism and significant number of metabolites, plasma concentration of moricizine has not demonstrated a correlation with antiarrhythmic efficacy.

Clinical Use

Moricizine has been evaluated for treatment in a variety of ventricular arrhythmias. It has been shown to be effective in suppressing VPCs and nonsustained ventricular tachycardia (107–109). The response to symptomatic nonsustained ventricular tachycardia, sustained ventricular tachycardia, and ventricular fibrillation have been variable. The overall response of these arrhythmias in patients undergoing PES is approximately 20%. The CAST study (CAST II) has reported an initial higher mortality in the moricizine-treated group than in the placebo-treated group, causing the study to be stopped (110).

Adverse Effects

Moricizine is well tolerated in the recommended doses. The adverse effects that have been observed are attributed to its phenothiazine structure and similar activity. The noncardiac adverse effects include dizziness, nausea, headaches, hypothesis, elevation of liver enzymes, and abdominal pain. Cardiac adverse effects include congestive heart failure, chest pain, and asymptomatic arrhythmias. The occurrence of proarrhythmic events has been reported to have an appropriate incidence of 3.2% with no relationship reported to dose. The malignant neuroleptic syndrome has been observed with moricizine (111).

Dosage and Administration

Moricizine is available in 200-mg, 250-mg, and 300-mg tablets. The recommended dosage ranges from 600 to 900 mg/day. Higher doses have been used. However, they have been associated with adverse events.

Class 2 Antiarrhythmic Agents

β-Adrenergic Blocking Agents

Numerous beta-blockers are presently available. Their clinical utility and effectiveness have been shown in a variety of cardiovascular, as well as noncardiovascular, conditions (112).

Pharmacology/Electrophysiology

The antiarrhythmic effects of the β-adrenergic receptor blocking drugs are chiefly related to their capacity to inhibit the β-receptor (113). Most of our information on the antiarrhythmic actions of beta-blockers come from studies with propranolol, although, to the extent they have been evaluated, other beta-blockers have produced similar results on the heart. Propranolol and some of the other beta-blockers have quinidine-like membrane-stabilizing effects that may play a small part in the antiarrhythmic effects of these drugs. However, in clinically relevant doses of beta-blockers, these membrane-active effects (Na$^+$ channel blockade) are minimal. Additional antiarrhythmic properties include slow conduction and increasing AV nodal refractoriness.

Pharmacokinetics

Many β-adrenergic blocking agents are available and their pharmacokinetics may favor one agent over another. Propranolol is frequently used and is considered to be the prototype beta-blocker. However, in the ICU setting where acute therapy is required, esmolol has been reported to be effective and possesses attractive pharmacokinetic properties (114): rapid onset of action and a short half-life. However, the volume in which the drug needs to be administered often severely limits its utilization.

Due to the wide variability in concentration among the beta-blockers, monitoring of serum concentration is usually not warranted. Rather, monitoring of heart rate and blood pressure are used to evaluate the degree of beta-blockade.

Clinical Use

The beta-blockers have been reported to be useful in the prevention and treatment of supraventricular arrhythmias, especially those due to the Wolff-Parkinson-White syndrome. These drugs are also effective in arrhythmias related to excessive cardiac adrenergic stimulation such as those associated with thyrotoxicosis, pheochromocytoma, exercise, emotion, and anesthesia with cyclopentane and halothane. The beta-blockers appear to be the drugs of choice for treating patients with ventricular arrhythmias associated with the prolonged QT interval syndrome (115). They are also effective in preventing exercise-induced augmented ventricular ectopy in patients with coronary disease, in treating arrhythmias associated with mitral valve prolapse syndrome, and in reducing the number and complexity of premature ventricular complexes after myocardial infarction. Several studies have demonstrated a reduction in the incidence of sudden death following myocardial infarction in patients chronically treated with β-adrenergic blockers such as practolol, propranolol, timolol, metoprolol, and atenolol. The reduction in sudden death is on the average of 25% in relative risk reduction. This translates into a significant reduction in sudden death after myocardial infarction. This reduction is related to the attenuation of sympathetic stimulation by the beta-blocking agents. Beta-blockers with intrinsic sympathomimetic activity (ISA), except in a small study with practolol, have not been shown to cause a reduction in sudden death in the postinfarction patients.

The beta-blocker may be very effective in the control of the ventricular response in atrial fibrillation. In patients with marked rate fluctuations in AF or in patients with significant pauses, pindolol, a beta-blocker with ISA, can be given with rate control and without the need for permanent pacing (116).

Adverse Effects

The most common adverse effects seen with beta-blockers are not cardiovascular (113). These include exacerbation of asthma and chronic obstructive pulmonary disease, fatigue, insomnia, impotence, and depression. Cardiac effects may include worsening of LV function with the development of congestive heart failure, sinus node dysfunction, AV block, claudication, or Raynaud's phenomenon.

Class 3 Antiarrhythmic Agents

Amiodarone

Initially introduced as an antianginal drug, amiodarone is a benzofuran derivative that was found early on to possess antiarrhythmic

action. A considerable literature has developed on this antiarrhythmic activity over the last 25 years. Amiodarone is an effective agent but because of significant toxicity, it has been reserved for patients unresponsive to other antiarrhythmic agents. Amiodarone was approved in the United States in 1988 for the treatment of life-threatening ventricular arrhythmias when other modalities of therapy failed.

Pharmacology/Electrophysiology

Amiodarone possess a variety of pharmacologic and electrophysiologic effects on the myocardium. Its principal effects include prolongation of the action potential duration in the atrium and ventricles, and prolongation of the effective refractory period in the atrium, ventricles, AV node, and His-Purkinje system (117). Amiodarone has been shown to increase the refractoriness of the accessory pathways both in the antegrade and retrograde direction. α- and β-antagonist properties for amiodarone have been identified. These effects are partially responsible for the slowing of the SA rate. Other effects include a calcium channel blocking action that may inhibit depolarization-induced automaticity and delayed afterdepolarization.

The ECG effects observed with amiodarone are prolongation of the PR and QT intervals. The most prominent and pharmacodynamically important action is QT prolongation. This action has correlated with amiodarone's antiarrhythmic efficacy (118). Measurement made by intracardiac catheters demonstrates prolongation in the AH and AV intervals.

Pharmacokinetics

Absorption of oral amiodarone is slow and variable. Peak concentrations are usually achieved 6.5 hours postingestion. The drug is widely distributed into body compartments including fat, muscle, liver, lung, and spleen. Amiodarone is 96% protein bound, primarily to albumin. The half-life of amiodarone has been estimated at 55 days while the elimination half-life of amiodarone may be as long as 100 days following chronic dosing (119). Plasma concentrations required for a pharmacologic action take a minimum of 1 to 3 weeks in the absence of a loading dose (120).

Amiodarone undergoes hepatic metabolism. The principal metabolite that has been identified in humans is desethylamiodarone. It possesses antiarrhythmic activity and has a longer half-life than amiodarone. Hepatic excretion in the bile is the principal route of elimination for both amiodarone and desethylamiodarone. Correlation between the plasma concentration of amiodarone and dose has been evaluated and plasma levels correlate with an effective range of drug action (121).

Clinical Use

Amiodarone has been reported effective in treating a wide variety of supraventricular (122) and ventricular arrhythmias (123,124). It is particularly effective in arrhythmias associated with the Wolff-Parkinson-White syndrome (125) and in the prevention of recurrent ventricular tachycardia and ventricular fibrillation. In many such cases, amiodarone is the sole agent to which patients with these difficult-to-treat arrhythmias have responded (126). Amiodarone has a high degree of efficacy for maintaining normal sinus rhythm in patients with atrial fibrillation or flutter, with rates of maintaining sinus rhythm of 87%, 70%, and 55% at 1, 3, and 5 years, respectively (127). In a resistant population, amiodarone is more effective than, and equally well tolerated as, flecainide (128). After 3 and 12 months, the group treated by amiodarone remained in sinus rhythm, 73% and 60% incidence, greater than quinidine, 70% and 50%, and flecainide, 49% and 34%, respectively. IV amiodarone was not much more effective than placebo in one acute study in converting AF to normal sinus rhythm (NSR) (amiodarone effective in 68% and placebo in 60%) (129).

Amiodarone is effective when administered intravenously in a loading dose. If oral loading is not employed, oral administration may take months to reach an effective steady-state level sufficient to provide antiarrhythmic action. Amiodarone's effects against atrial arrhythmias and those of the Wolff-Parkinson-White syndrome may require a lower dose than what is needed in the treatment of the most severe ventricular arrhythmias. A report by Horowitz and colleagues (130) has shown amiodarone to be guided effectively by electrophysiologic testing. Additionally, in a randomized study comparing amiodarone therapy to the success of the implantable defibrillator, Gottlieb and colleagues (131) found them to be equally effective. However, in a larger, more recent study, the implantable defibrillator in the first few years is superior to amiodarone therapy (101). However, when amiodarone is compared to conventional therapy in patients with an implantable defibrillator, the amiodarone treatment group shows a better survival free of cardiac death, resuscitated VF, or syncopal defibrillator shock: amiodarone 82% vs 60% conventional at 2 years, amiodarone 69% vs conventional 52% at 4 years, and amiodarone 53% vs conventional 40% at 6 years (132).

Perhaps the area that amiodarone is studied most thoroughly is in the setting of high-risk patients post-MI. This is an especially useful area since the CAST trial has shown the adversity that the type Ic sodium channel blockers can cause in this high-risk population. The Basal Antiarrhythmic Study of Infarct Survival (BASIS) study looked at patients post-MI with asymptomatic complex ventricular arrhythmias and reported low-dose amiodarone to be superior to individualized antiarrhythmia treatment or no antiarrhythmia treatment (133). A large secondary prevention trial with amiodarone at a low dose (200–400 mg/day) was reported to reduce cardiac mortality by 42% (136). However, the mortality in both the control and amiodarone-treated group was substantially higher than what we expect post-MI in the United States. These trials were supportive of the thesis that amiodarone was indeed an effective agent in reducing sudden death in the post-MI population. Two larger trials were undertaken with amiodarone. The European Myocardial Infarct Amiodarone Trial (EMIAT) evaluated the efficacy of amiodarone on mortality of patients with decreased left ventricular function following MI with greater than 700 patients enrolled (135). A recent report of the findings reported no difference in total mortality as analyzed by the intent to treat analysis between amiodarone and placebo therapy (136). There was a trend toward reduction in sudden death mortality. Unfortunately, a considerable number of patients had stopped amiodarone therapy and this is a bothersome problem since many in the analysis were not receiving antiarrhythmic therapy.

The Canadian Amiodarone Myocardial Infarction Arrhythmia Trial (CAMIAT) randomized survivors of acute myocardial infarction (AMI) with complex ventricular premature depolarization to amiodarone or placebo (137). A report of the findings at the American Heart Association meeting showed no significant difference in relative or absolute risk reduction, though a trend did favor amiodarone (138). There was a reduction in sudden death by amiodarone, and CAMIAT patients with previous MI or who had the complication of pulmonary edema benefited the greatest from amiodarone with greater than four events prevented per 100 patient years of treatment. Unlike the Class 1c agents, amiodarone, despite its toxicities, does not cause more harm than benefit and may reduce electrical instability. The findings suggest that alternatives in the future to amiodarone without noncardiac toxicities are needed.

Patients with congestive heart failure are also at high risk for sudden arrhythmic death (139,140). Retrospective studies by Cleland et al. (141) and by Chatterjee (142) have reported reduced mortality in patients with heart failure treated with amiodarone. Nicklas and associates (143), in a small but randomized prospective trial, reported no reduction in mortality in CHF patients on amiodarone. A large randomized study mostly in patients with nonis-

chemic myopathy, the Grupo de Estudio de la Sobrevida en la Insuficiencia Cardiaca en Argentina (GESICA), reported a 28% reduction in overall mortality in the amiodarone-treated group as compared to control, a highly significant finding (144). Sudden death was reduced by 27% and death due to progressive heart failure was reduced by 23%. The Congestive Heart Failure: Survival Trial of Antiarrhythmic Therapy (CHF-STAT) study of amiodarone in patients with congestive heart failure and asymptomatic ventricular arrhythmia reported no significant difference in mortality between the amiodarone- or placebo-treated group (145). At 2 years, the incidence of sudden death was 15% in the amiodarone group and 19% in the placebo group ($p = .43$). There was a trend toward reduction in mortality among patients with nonischemic cardiomyopathy who received amiodarone ($p = .07$). Amiodarone was effective in suppressing ventricular arrhythmias and increased left ventricular ejection fraction by 42% at 2 years. One difference between the GESICA study and the VA study is that 70% of GESICA patients had nonischemic myopathy and only 39% of the VA study patients had a nonischemic myopathy. In fact, in the GESICA study, the nonischemic group was the subgroup in which improvement was noted. In addition, a substantial number of patients discontinued amiodarone in the VA study, 27%, due to intolerable side effects. Thus, by using intent to treat analysis, a quarter of the evaluated patients were not on antiarrhythmic therapy. The high rate of discontinuation is a persistent problem making intent to treat analysis evaluation of antiarrhythmic studies problematic. However, the high rate of discontinuation should not be attributed to side effects alone since in the VA amiodarone study 23% of those on placebo discontinued due to intolerable side effects. Further confusing matters is that an additional 46% of patients on amiodarone and 32% on placebo withdrew from the study or were lost to follow-up. With the study thus evaluating only about 50% of those initially starting therapy, the results are unfortunately far from definitive.

IV Amiodarone

For several decades, the treatment of acute arrhythmias has centered around IV lidocaine and bretylium. However, the availability of IV amiodarone, a recent development in the United States, has created a viable alternative leading to a review of the IV antiarrhythmic field and particularly the role of IV amiodarone. The standard therapy of lidocaine offers a limited efficacy. One report by Pacifico and associates (49) finds an 8% efficacy and a second report by Gorgels et al. (146) reports an efficacy of 21% as compared to IV procainamide of 80% (in a drug-responsive population). Amiodarone has been used for years IV in Europe on the basis of empirical data. The usual dose is 150 mg IV followed by a second ampule as needed. The infusion must be slow since the formulation used may cause significant hypotension and negative inotropy. Munoz and associates (147) reported that amiodarone without Tween 80 causes less hypotension, though a slow infusion of the amiodarone with Tween 80 can avoid this problem. Helmy and associates (148) reported on 46 patients with recurrent drug-refractory sustained ventricular tachycardia or ventricular fibrillation or both who showed a 59% response rate to IV amiodarone. Kowey and associates (149) found that patients who responded acutely to IV amiodarone with incessant VT showed a good late response to oral amiodarone and that predischarge electrophysiology testing even in patients who have polymorphous VT had predictive value over and above the observed clinical response.

Recently, two large studies have shown the acute efficacy of IV amiodarone in patients with resistant arrhythmias. Levine and associates (150) reported on 273 patients with recurrent VT refractory to lidocaine and bretylium who received one of three doses of IV amiodarone (though in the end all patients received about the same concentration) with a 40% response rate to amiodarone. A significant difference in time to first recurrence of VT over the first 12 hours was observed. A double-blind comparison of IV amiodarone with bretylium in patients with recurrent hemodynamically destabilizing ventricular tachycardia or fibrillation in 302 patients showed comparable efficacy between bretylium and amiodarone (151). Significantly more patients treated with bretylium had hypotension compared with the two amiodarone groups. Sheinman and associates (152) reported a dose-ranging study with 342 patients showing that following amiodarone there was a significant dose-related increase in the time to first event and a significant dose-related decrease in the number of supplemental boluses per hour. Hypotension (systolic <80 mm) was the most common adverse emergency event (26%). This high incidence of hypotension was seen also in the Levine report with a 6% mortality attributed to hypotension. Thus, IV amiodarone is given in a slow infusion over 10 minutes to avoid life-threatening hypotension. The hypotension is probably secondary to the myocardial depressant and vasodilating properties of benzyl alcohol and Tween 80, which are used to solubilize amiodarone in the Cordarone preparation.

A new amiodarone IV preparation, Amio Aqueous, is currently in clinical studies. It is devoid of Tween 80 and benzyl alcohol, and it solubilizes the amiodarone in an aqueous buffer. The recommended 10-minute slow infusion for Cordarone is not needed with Amio Aqueous. In animal studies, less hypotension and negative inotropy is found with Amio Aqueous (153,154). A recently completed clinical study in man reports no hypotension or negative inotropy following IV boluses of 150 mg of Amio Aqueous (155). There is considerable potential in the administration of amiodarone by rapid IV bolus, and this rapid means of administration should place amiodarone ahead of lidocaine as the first-line pharmacologic therapy for VT. Lidocaine is no safer and the literature appears to show it to be much less effective than IV amiodarone. Also, IV amiodarone may be useful for loading, decreasing hospitalization time (156).

Treatment of Atrial Arrhythmias

Amiodarone's use in the treatment of SVT is evolving in light of concerns regarding the adverse profile of amiodarone but the low reported proarrhythmia. However, amiodarone does demonstrate efficacy along with the potential of a reduced toxicity profile when employing a lower dose. What is especially attractive is the low proarrhythmic profile of amiodarone, especially in light of concerns raised with quinidine, sotalol, and flecainide. Amiodarone has been reported to be effective in controlling the rate of the ventricular response in atrial fibrillation (129). Amiodarone's effectiveness is at rest as well as at activity and amiodarone may be effective when other agents are unsuccessful (157). Amiodarone is effective in difficult-to-control paroxysmal atrial tachycardias and preexcitation-mediated tachycardia (158). Amiodarone may be used to reduce the episodes of paroxysmal AF and the recurrence rate of AF following cardioversion (127,159). In fact, in patients with refractory AF and flutter, amiodarone showed an efficacy rate of 86% over 2.5 years in maintaining sinus rhythm (127). Still, the rates of withdrawal because of adverse effects were 0.08, 0.22, and 0.30 at 1, 3, and 5 years, respectively. However, no deaths were reported in the study, further supporting the proarrhythmic safety profile of amiodarone. One caution, however, is that amiodarone has at times been reported to convert atrial fibrillation to atrial flutter, which may accelerate the ventricular response by decreasing concealed conduction even with the AV node conduction-blocking effects of amiodarone (160).

Adverse Effects

In antiarrhythmic doses, chronic amiodarone therapy has little effect on myocardial contractility. However, there have been cases of IV depression associated with amiodarone especially following IV administration due to the solubilizers Tween 80 and benzyl alcohol.

Corneal microdeposits are reported in most patients following chronic administration. Rash, skin discoloration, photosensitivity, thyroid dysfunction, and serious neurologic side effects, including peripheral neuropathy, constipation, and gastrointestinal symptoms, are but a few of the side effects of amiodarone. The most serious adverse actions are hepatic, renal, and pulmonary. Pulmonary fibrosis, which can be severe and cause death, occurs in about 15% of patients receiving very high, chronic oral doses of amiodarone (126,161).

Dosage and Administration

To avoid the gastrointestinal side effects, the drug is usually administered in two or three doses daily. One dosing regimen used is 800 to 1600 mg daily for 2 to 3 weeks followed by a maintenance dose of 600 to 800 mg/day for 4 weeks and then 200 to 400 mg/day thereafter. Many experienced in the use of amiodarone are turning to low-dose approaches. The European experience has been that lower doses may be associated with a better complication rate and an adequate efficacy profile. Studies are needed to confirm that doses of 100 to 200 mg/day may be as effective as the higher doses previously employed. Further reduction in amiodarone doses are needed in patients on amiodarone for 1 year or more because drug concentration accumulates. Monitoring drug levels and keeping the amiodarone and its metabolite desethyl amiodarone between 1.5 and 2 Ng/ml may avoid toxicity. Additionally, monitoring the QT or JT prolongation may be a physiologic indicator of drug effect (118). Another study has suggested that decreasing QT dispersion may be an indicator of amiodarone efficacy (162).

Bretylium

Bretylium was first introduced in the 1950s as an antihypertensive agent, and was subsequently discontinued due primarily to the adverse effect of orthostatic hypotension. Further work by Bacaner subsequently revealed significant antifibrillatory and antiarrhythmic activity. It was reintroduced for this use in 1978 and was approved for intravenous administration in the treatment of life-threatening ventricular arrhythmias.

Pharmacology/Electrophysiology

Bretylium is a bromobenzyl quaternary ammonium structure. The antiarrhythmic actions of bretylium are partly a result of its indirect action by interfering with the adrenergic nervous systems neural transmission, and partly a direct action on myocardial ion transport (163). The acute electrophysiologic changes of bretylium are complex and result from the reflex autonomic activation due to vasodilation as well as a result of its adrenergic neuronal effects. These actions increase automaticity of the sinus node, Purkinje fibers, conduction velocity, and V_{max}. The action potential amplitude is also increased, while the action potential duration and effective refractory period are decreased.

Direct effects of bretylium after prolonged administration, seen on the myocardium, include an increase in action potential duration and the effective refractory period in the Purkinje fibers and ventricular muscle. Bretylium possesses significant antifibrillatory properties, making sustained VF induction more difficult and decreasing the myocardium's capacity to sustain VF. The drug also may spontaneously terminate ventricular fibrillation through catecholamine release and blockade of reuptake. The acute catecholamine release causes an adrenergic storm behaving not too dissimilar from the depolarization caused by electroshock. Bretylium may also reduce energy requirements for successful cardiac defibrillation.

Pharmacokinetics

Bretylium is available in a parenteral formulation. The intramuscular administration of bretylium has been used, although its onset is slower with variable absorption. Bretylium is extensively distributed into adrenergic nerves. The volume of distribution of bretylium ranges from 3 to 5 L/kg. Bretylium is not significantly protein bound and is primarily cleared via the kidneys. The terminal half-life is long and ranges from 7 to 13 hours after loading and prolonged administration (164). The clearance of bretylium has been correlated with creatinine clearance in patients with renal dysfunction. Therapeutic plasma monitoring is not of value in predicting therapeutic efficacy or toxicity (165).

Clinical Use

Bretylium is indicated for the treatment of sustained ventricular tachycardias and/or ventricular fibrillation. It has been used in the drug selection process primarily when lidocaine fails. In the setting of a ventricular fibrillation arrest, bretylium appears to have no greater efficacy that lidocaine (166). Additionally, bretylium may impair hemodynamic recovery following arrest as demonstrated in animals (167). Peak antiarrhythmic effects (class III action) are commonly observed several hours following administration, contrary to its antifibrillatory effects, which are seen within minutes.

Adverse Effects

Initial observations seen with bretylium are an increase in blood pressure and heart rate followed by hypotension, as a result of its release of norepinephrine from nerve endings and the subsequent blockade of reuptake and release. Hypotension causes discontinuation of therapy in approximately 10% to 30% of patients. The use of protriptyline, a tricyclic antidepressant, 5 to 10 mg every 6 to 8 hours has been found effective in preventing the orthostatic hypotension that is observed following bretylium administration (168). The mechanism of action of protriptyline or nortriptyline is to block bretylium uptake into the sympathetic ganglions critical in maintaining postural blood pressure. Nausea, vomiting, and retching have been reported with intravenous administration. Additional adverse reactions include diarrhea, flushing, anxiety, dyspnea, diaphranesis, and nasal stuffiness.

Dosage and Administration

Bretylium is administered as a 5- to 10-mg/kg bolus injection for the treatment of unstable ventricular tachycardia or ventricular fibrillation. Maintenance therapy may be continued using bolus dosing of 5 to 10 mg/kg every 6 to 8 hours or as a continuous infusion of 1.0 to 2.0 mg/min.

Sotalol

Sotalol is a β-adrenergic blocking agent that has been available in Europe for years for hypertension treatment. Recent studies have limited its indication to arrhythmia therapy, as it is used in the United States. Sotalol possesses both Class 2 and Class 3 antiarrhythmic properties. It was approved by the FDA in 1992 for the treatment of life-threatening ventricular arrhythmias.

Pharmacology/Electrophysiology

Sotalol is a nonselective β-adrenergic blocking agent. In contrast to other beta-blockers, its chemical structure is a methane-sufulonamide-substituted phentolamine. It is devoid of both intrinsic sympathomimetic activity and local anesthetic activity. Sotalol has approximately one-third the beta-blocking activity of propranolol.

Sotalol possesses potent electrophysiologic effects not observed with conventional beta-blockers. These effects include prolongation of the effective refractory period in both the atria, ventricles, atrial ventricular node, and bypass tracts (169). Sotalol decreases the resting heart rate. Hemodynamically sotalol causes a reduction in the cardiac output without changing stroke volume in hypertensive patients with CHF. However, in the normotensive patient without CHF, cardiac output is not altered. Electrocardiographically sotalol increases the QTc and JT intervals but does not affect the QRS duration.

Pharmacokinetics

Sotalol approaches nearly 100% absorption. Following oral administration peak plasma concentration is achieved within 3 hours. Sotalol undergoes no hepatic metabolism and is principally excreted by the kidneys through glomerular filtration. Both plasma clearance and renal clearance have correlated with creatinine clearance (170). The elimination half-life ranges from 7 to 18 hours in patients with normal renal function and increases with renal dysfunction.

Clinical Efficacy

Sotalol has been evaluated in a variety of supraventricular arrhythmias. Sotalol has been shown to be effective in terminating proximal supraventricular tachycardia. In a series of studies IV sotalol converted atrial flutter and atrial fibrillation in 62% and 27%, respectively. Sotalol prevented inducible supraventricular tachycardia in 59% of patients compared with 28% with metoprolol (171). Comparison of sotalol with digoxin-quinidine for conversion of acute atrial fibrillation to sinus rhythm (the Sotalol-Digoxin-Quinidine Trial) was undertaken by Halinen and colleagues (172). Conversion of AF to sinus rhythm occurred in 17 of 33 patients (52%) taking sotalol and 24 of 28 patients (86%) taking quinidine. Treatment was discontinued in 16 patients taking sotalol (48%) because of asymptomatic bradycardia or hypotension, and in 20 patients taking quinidine (71%). Asymptomatic wide complex tachycardia (QRS >0.12 sec) was found in 13% and 27% of patients taking sotalol and quinidine, respectively. The author concludes, "Both quinidine and sotalol are potentially dangerous and cannot be recommended for self-medication for acute AF outside of hospital" (172). In another study, propafenone was compared to sotalol for suppression of recurrent symptomatic atrial fibrillation with 50 patients randomized to sotalol and 50 to propafenone (173). The proportion of patients remaining in sinus rhythm was similar between propafenone 46 ± 8, 41 ± 8, and 30 ± 8 as compared to sotalol 49 ± 7, 46 ± 8, and 37 ± 8 at 3, 6, and 12 months, respectively.

Ventricular Arrhythmias

In the treatment of ventricular arrhythmias sotalol appears to be effective in suppressing premature ventricular contractions (primarily higher grades of ventricular ectopy). In two separate electrophysiology studies IV sotalol prevented VT/VF reinducibility in 46% and 67%, respectively (174,175). Sotalol prevents ventricular tachycardia induction in 33% of patients as contrasted to a procainamide efficacy rate of 22% (176). Mason and the ESVEM investigators (177) reported on a comparison of several antiarrhythmic drugs in patients with ventricular tachycardia. Drugs were guided by either electrophysiologic (EP) testing or Holter monitoring. Sotalol was more frequently effective in the EP-guided group but there was no difference in the Holter group in terms of acute efficacy. The percentage of patients with adverse drug effects was lowest among those receiving sotalol. The actual probability of a recurrence of arrhythmia after a predictor of drug efficacy by either strategy was significantly lower for patients treated with sotalol than for patients treated with the other drugs. Sotalol reduced risk of death from any cause (risk ratio 0.50), death from cardiac causes (risk ratio 0.50), and death from arrhythmia (risk ratio 0.50, $p = .04$). Those patients on sotalol showed a higher incidence of persistent drug effectiveness than for any other antiarrhythmic drug.

Sotalol is a racemic mixture of d,l-isomers with the beta-blocking properties ascribed mostly to the l-isomer and the class III activity mostly to the d-isomer (178). A promising thesis was to utilize the class III agent d-sotalol since the racemic dl sotalol caused depression of LV function and impaired airway function like other beta-blockers. The effectiveness of d-sotalol was evaluated in a high-risk population post-MI with an ejection fraction less than 40%. The Survival with Oral-d Sotolol or SWORD study was stopped prematurely because of a higher mortality in the d-sotalol treated group. The percent cumulative mortality after about 350 days from randomization was 8.9% for sotalol and 7.3% on placebo (179). The percentage cumulative probability of reinfarction was 3.5% on placebo and 5.9% on sotalol. These results were surprising especially in light of an earlier study by Julian et al. (180) in which d,l-sotalol shows a reduced cumulative mortality (7.3%) than for placebo (8.9%) after 1 year. These findings support the possible importance of the Class 2 component in sotalol and possibly in amiodarone as well. However, the incidence of torsades de pointes on sotalol can range from 6% to 15% or higher in some special populations (181,182), while the incidence of torsades de pointes due to amiodarone is exceedingly rare (183). Interestingly, sotalol shows a gender difference with an increase in torsades de pointes in females (184). These observations are similar to those reported with a number of other agents that prolonged APD and are associated with torsades de pointes VT (185).

Adverse Effects

Similar to other beta-blocking agents, sotalol has dramatic cardiac and noncardiac side effects. These include bradycardia, hypotension, dyspnea, dizziness, worsening CHF, fatigue, and impotence and asthma exacerbation. As the dose of sotalol increases above 320 mg, the Class 3 effect increases; below 320 mg/day the beta-blocking effects are predominant. Sotalol causes marked QT (JT) prolongation, and thus close monitoring for torsades de pointes arrhythmia, which can be worsened by hypokalemia or hypomagnesemia, is appropriate. The occurrence of proarrhythmia ranges between 5% and 20%. For these reasons, it is recommended that initiation of sotalol be done in the hospital. However, the observed proarrhythmia in the SWORD study was cumulative, not all occurring at the initiation of therapy, similar to the CAST results.

Dosage/Administration

Sotalol is available in 50-, 100-, and 200-mg tablets. The daily dose ranges from 100 to 600 mg, though doses in considerable excess have been employed in arrhythmia trials. The prescribing information goes up to 640 mg/day with a table relating increased dose to a mean increase in QTc and the incidence of torsades de pointes ventricular tachycardia. These relationships between torsades and QTc increments appear to be dose related.

Ibutilide

Ibutilide is a recently introduced agent for the conversion of AF to sinus rhythm. The mechanism of action of ibutilide remains controversial. In freshly isolated adult heart cells using standard voltage clamp techniques, Lee and Gibson (186) found that the inhibition of Ikr by ibutilide plays a minor role and the increased duration of the action potential is caused by activation of I_{Na}-S channel. Young et al. (187,188) reported that in atrial tumor (AT-1) cells ibutilide is an inhibitor of Ikr. Regardless of the current inhibited, the predominant electrophysiologic action of ibutilide is to prolong the action potential duration.

Clinical Action

Stamble and associates (189) evaluated IV ibutilide for rapid conversion of atrial flutter or fibrillation in 266 patients (133 AF and 133 with atrial flutter). The conversion rate was 47% after ibutilide and 2% after placebo. The success of conversion was not dose related. The efficacy was higher in patients with atrial flutter than fibrillation (63% vs 31%). Arrhythmic termination occurred in a mean of 27 minutes after infusion was started. Proarrhythmia was noted in 15 out of the 180 ibutilide-treated patients, an 8.3% incidence of polymorphic VT. Ellenbogen et al.'s (190) dose-response study evaluated ibutilide efficacy in AF and atrial flutter termination. Atrial fibrillation was terminated in 34%: 3% of the placebo treated, and 12%, 33%, 45%, and 46%, respectively, of the patients treated with 0.005, 0.010, 0.015, and 0.025 mg/kg ibutilide. The mean time to arrhythmic termination was 19 minutes (range 3 to 70 minutes). In this study, successful arrhythmia termination was not affected by enlarged left atrial diameter, decreased ejection function, presence of valvular heart disease, or the use of concomitant medications (beta-blockers, Ca^{2+} blockers, or digoxin). Polymorphic VT was noted in approximately 4% of patients. In a third study, ibutilide converted atrial flutter in 14 of 25 patients (56%) in a mean time of 25 ± 16 minutes (191). Conversion of atrial flutter by ibutilide was characterized by increased variability in atrial cycle length and diastolic interval.

Adverse Drug Reactions

Perception of stroke on conversion has been reported, and appropriate anticoagulation applying the same guidelines as with cardioversion is appropriate. Suspension of LV function and CHF has been reported as well as exacerbation of underlying condition disturbance. Proarrhythmia with the development of torsades de pointes ventricular tachycardia is the most frequent and serious side effect reported with ibutilide. The incidence varies from 3% to 15%. The majority of reported cases from a number of studies are in patients with significant LV dysfunction. It would seem logical that hypokalemia or hypomagnesemia would facilitate torsades. A further breakdown of the proarrhythmia was 3.8% among men, 13.2% among women, 15.9% among nonwhites, 3.6% among white males, 11.4% among patients with heart failure, and 3.6% among patients without heart failure.

Dosage and Administration

Ibutilide is administered by IV infusion at doses of 1 mg over 10 minutes followed by 0.5 mg 10 minutes later over 10 minutes, or 1 mg followed by 1.0 mg ibutilide, each infusion over 10 minutes. Doses for patients weighing under <60 kg are adjusted, 1.0 mg = 0.01 mg/kg, 0.5 mg = 0.005 mg/kg.

Drug Interactions

Patients should be normokalemic and eumagnesemic. Ibutilide should not be combined with other agents known to cause torsades de pointes tachycardia like terfenadine or with other antiarrhythmic agents known to prolong the action potential duration: quinidine, sotalol, or amiodarone.

Class 4 Calcium Channel Blockers

Verapamil

Verapamil was the first calcium channel blocker developed that inhibits the transmembrane flux of calcium ions in excitable tissue. Its use is recommended in the treatment of angina, hypertrophic cardiomyopathy, hypertension, and supraventricular arrhythmias.

Pharmacology/Electrophysiology

Verapamil, a diphenylalkamine, is structurally different from other calcium channel blocking agents. It exerts its antiarrhythmic effect by blocking the transmembrane movement of calcium ions in myocardial tissue (192). Electrophysiologically, the greatest effects of verapamil are observed in the SA and AV nodes, where verapamil decreases automaticity and conduction. Verapamil slows conduction and has peripheral vasodilatory actions. Verapamil has lesser effects on the atrial tissue and minimal effects on ventricular and Purkinje fibers. Verapamil prolongs the AH interval and the effective refractory period of the AV node. On the surface ECG, verapamil prolongs the PR interval.

Pharmacokinetics

Following oral administration, verapamil is rapidly absorbed. Peak serum concentrations are usually achieved within 90 minutes (193). Verapamil undergoes significant first-pass metabolism and thus only 10% to 20% of the drug is bioavailable. In patients with liver cirrhosis or hepatic dysfunction, bioavailability approaches 100%. The volume of distribution ranges from 3 to 6 L/kg and the protein binding is approximately 90%. Verapamil undergoes extensive hepatic metabolism. The principal metabolite is norverapamil. It possesses approximately 20% of the calcium channel blocking activity of verapamil. The elimination half-life of verapamil is 4 to 5 hours following acute dosing. However, following chronic administration, nonlinear accumulation occurs and the half-life increases to 8 to 12 hours. Plasma concentrations are not routinely used; however, concentrations range from 80 to 300 ng/ml.

Clinical Use

Intravenous verapamil is very effective in the treatment of supraventricular arrhythmias (194,195). Verapamil results in a clinically significant reduction in the ventricular response in atrial fibrillation or flutter and can convert paroxysmal supraventricular tachycardia to sinus rhythm (196). Verapamil has not been effective in reducing the frequency of ventricular ectopic beats or in reducing ventricular arrhythmia frequency (197), although some authors have reported variable degrees of success (198,199). Verapamil is not effective in preventing ventricular tachycardia induction in the PES laboratory. In patients with normal LV function with verapamil-responsive ventricular tachycardia, the drug is effective. The efficacy of verapamil in preventing induction of ventricular tachycardia using electrophysiology techniques has been observed to be considerably less than that seen with procainamide (200). The lack of ectopy suppression combined with the prevention of ventricular tachycardia induction in some patients supports the thesis that triggering mechanism and threshold for inducibility may be affected differently by the same drug. The long-term clinical efficacy of verapamil in preventing recurrence of ventricular tachycardia and ventricular fibrillation remains to be determined. Certainly the anti-ischemic action of the drug may be a contributing factor to its antiarrhythmic effects and this attribute will have to be taken into consideration while the drug is used in clinical practice. Verapamil may be especially effective in the control of the ventricular response in atrial fibrillation in the ambulatory patient. While digoxin may slow the ventricular response at rest, it is not effective during exercise; however, verapamil is effective (201,202).

Adverse Effects

The most commonly observed adverse effect with verapamil is transient lowering of the blood pressure. This occurs following intravenous administration. Preadministration of calcium salts has been shown to be beneficial in preventing hypotension and may not alter the antiarrhythmic activity (203). Bradycardia or AV block has occurred primarily in patients concomitantly receiving beta-blocker therapy. Verapamil is not recommended in patients with sick sinus syndrome, second- or third-degree AV block, or in patients with left ventricular ejection fraction less than 30%. Other adverse effects that may occur following chronic dosing include constipation, headache, dizziness, and peripheral edema. Verapamil should not be used in patients with

a possible preexcitation syndrome since the AV node may be blocked, permitting rapid conduction of the atrial depolarizations (especially when patient is in AF) to the ventricle by way of the accessory pathway, causing hemodynamic collapse (204).

Recently, questions have arisen regarding the use of calcium channel blockers and the possibility of their contributing to an adverse outcome (205). The relation of these findings to verapamil is very tenuous. Verapamil in both *in vivo* and *in vitro* experiments reduces ischemic injury and infarct size (206,207). Pretreatment with verapamil prevents postischemic stunning (208). The Danish Verapamil Infarction Trial (DAVIT I) showed that verapamil was effective in preventing reinfarction. Retrospective analysis demonstrated a lower mortality in the verapamil group, 3.7% compared with 6.4% in the placebo group (p = .05) (209). In DAVIT II 878 patients started treatment with verapamil 360 mg/day and 897 patients on placebo (210). Treatment started the second week after infarction and continued for up to 18 months. The findings were 95 deaths and 146 major events in the verapamil group, and 119 deaths and 180 major events in the placebo group. In patients without heart failure mortality ratio was 7.7% on verapamil and 11.8% on placebo. In patients with heart failure mortality was 17.9% and 17.5% on or off verapamil. Thus, in a high-risk population, which in CAST and SWORD revealed increased risk for those on the pharmacologic agents studied, verapamil posed no risk. Verapamil should not be used to prolong life post-MI since this has not been found with post-MI therapy. But in patients with SVT, AF (for rate control), or responsive VT, verapamil therapy would not be deleterious.

Dosage and Administration

Verapamil is available both in the oral and intravenous formulations. The recommended oral dose ranges from 120 to 480 mg/day. A sustained release formulation is available in 120-, 180-, 240-, and 300-mg doses. The recommended intravenous dose is 0.025 to 0.15

mg/kg or 5 to 10 mg over 1 to 3 minutes. If optimal response is not achieved, a second bolus may be administered after 5 to 10 minutes. Alternatively, a continuous infusion of 5 mg/hour has been demonstrated to be effective in diminishing the hemodynamic fluctuations seen with repeated boluses (211).

Diltiazem

Diltiazem is a calcium channel blocking agent that has been extensively used in the treatment of angina and hypertension. Although it has been known to have antiarrhythmic activity, it use has been limited. The recent development of the intravenous formulation has allowed it to be used in the acute setting for the treatment of SVT. The drug has been approved for SVT therapy by the FDA in 1991.

Pharmacology/Electrophysiology

Diltiazem is a benzothiazepine derivative. Its mechanism of action is comparable to other calcium channel blocking through inhibition of slow inward current of calcium in voltage-dependent channels. Electrophysiologic studies have shown that diltiazem slows AV nodal conduction and prolongs AV nodal refractoriness (212). Diltiazem has no significant effect on refractory periods of the atrial, ventricular, or Purkinje fibers. In comparison to verapamil, diltiazem has less of an effect on the AV node, but greater vasodilatory properties. Diltiazem has less of an effect in suppressing left ventricular function than verapamil. Thus, diltiazem may have an advantage in preserving left ventricular function in patients with SVT and impaired LV function. On the surface ECG, diltiazem prolongs the PR interval. The safety and efficacy of IV diltiazem were evaluated in patients with atrial flutter and atrial fibrillation (213). Overall, 94% of patients responded to diltiazem with a 20% or greater reduction in blood pressure. Seventy-eight patients received an infusion; 47% maintained response with 5 mg/hr, 68%

maintained response after increasing the infusion to 10 mg/hr, and 76% after titration of the infusion to 15 mg/hr. Hypotension was the most common side effect, occurring in 13% of patients; 3.6% were symptomatic.

Pharmacokinetics

Diltiazem is 90% absorbed following oral administration. It undergoes rapid first-pass metabolism to form several metabolites. *N*-mono-desmethy diltiazem is the principal metabolite and it possesses 20% of diltiazem activity. The mean oral bioavailability is 30% to 40%. Diltiazem is 80% to 90% protein bound to albumin. The elimination half-life of diltiazem ranges from 2 to 11 hours following oral administration and 2 to 5 hours following intravenous administration. The pharmacokinetics of diltiazem are not altered in patients with renal dysfunction. However, the pharmacokinetics of diltiazem are altered in patients with hepatic dysfunction or impaired hepatic clearance.

Clinical Efficacy

Diltiazem has been shown to be effective clinically in the treatment of supraventricular tachycardia (214). Oral diltiazem has been shown to be effective in the treatment of PSVT with an accessory pathway (i.e., Wolff-Parkinson-White syndrome). However, diltiazem should be used with caution since the same pharmacologic adverse action with verapamil could occur with diltiazem as well (215). Recently, with the availability of the intravenous formulation, diltiazem has been demonstrated to be safe and effective in the acute treatment of atrial fibrillation/flutter, administered as a bolus or as a continuous infusion (216).

Adverse Effects

Diltiazem is well tolerated by patients receiving either the oral or the intravenous formulation. Adverse effects that may be observed include headache, flushing, hypotension, nausea, dryness of the mouth, and constipation. Significant AV block may occur, although it is principally seen in patients receiving concomitant medication (i.e., beta-blockers).

The effects of diltiazem on outcome are important in evaluating the agent for long-term use in cardiac patients. The effect of diltiazem on long-term outcome after acute MI was assessed in 2,377 patients enrolled in the Multicenter Diltiazem Post-Infarction Trial and subsequently followed for 25 ± 8 month. Among the patients with first non–Q-wave and first inferior Q wave MI, there were fewer cardiac events during follow-up in the diltiazem than in the placebo group. There were more events in the diltiazem-treated group in patients with an anterior Q-wave MI and a history of prior infarction (217). The different outcomes correlated with LV function, with a poorer outcome on diltiazem occurring in patients with poorer LV function. An earlier study also reported a benefit (preventing early reinfarction and severe angina) of diltiazem in patients with a non–Q-wave infarction (218).

Dosage and Administration

Oral dosing of diltiazem ranges from 120 to 360 mg/day. The recommended dose of the intravenous formulation, which is approved for supraventricular tachycardia, is 0.25 mg/kg over 2 minutes. If the response is inadequate, the patients may be rebolused with 0.35 mg/kg over 2 minutes, followed by a maintenance infusion 5 to 15 mg/hr.

Adenosine

Adenosine is a naturally occurring compound that has been approved for the acute treatment of supraventricular arrhythmias.

Pharmacology/Electrophysiology

Adenosine is an endogenous purine nucleoside that has been described to have a variety of physiologic effects. Its activity on the myocardium is through stimulation of adenosine 3′,5′-cyclic monophosphate (cAMP) and the adenosine receptor. Two adenosine receptors (A_1 and A_2) have been identified. Adenosine has a number of significant physiologic actions. Adenosine causes both arterial vasodilation as well as coronary vasodilation (219). The principal electrophysiologic effect of adenosine is prolongation of AV refractoriness and conduction. Adenosine also causes a slowing of the heart rate by directly decreasing SA node automaticity. In electrophysiologic studies adenosine demonstrates a dose-dependent increase in AH conduction with AV block.

Pharmacokinetics

Adenosine is administered intravenously. In plasma, adenosine is rapidly taken up by the erythrocytes in the endothelial system. The drug is rapidly metabolized in the circulation resulting in an ultrashort half-life, ranging from 0.6 to 1.5 seconds and a duration of effect lasting approximately 1 to 2 minutes.

Clinical Use

Adenosine has been shown to be effective in the termination of proximal supraventricular tachycardia (220). Rarely, it may terminate an adenosine responsive ventricular tachycardia (221). Adenosine IV has been used with success in the treatment of AV nodal tachycardia and Wolff-Parkinson-White syndrome tachycardias using the AV node as part of the reentry circuit (222). It holds additional benefit in the diagnosis of wide complex tachycardia versus supraventricular tachycardia. However, the use of adenosine with ventricular tachycardia has resulted in profound hypotension causing cardiac collapse and arrest. The sensitivity and specificity of adenosine in diagnosing a wide complex tachycardia versus supraventricular tachycardia are high (223).

Adverse Effects

The most frequent adverse effects observed with intravenous adenosine include, facial flushing, dyspnea, nausea, light-headedness, dizziness, and syncope. Hypotension and bradycardia may occur transiently. Acute treatment with aminophylline has been shown to block these effects. Conversely, dipyridamole has been demonstrated to potentiate adenosine's action, and their combined use is to be avoided.

Dosage and Administration

The recommended dose for treatment of SVT is 6 mg over 1 to 2 minutes. A second dose may be administered if the desired effect has not been achieved. Alternatively, an incremental dosing regimen has also been suggested. Administration of 1 to 3 mg initially is followed by 2.5 to 3.0 mg until a maximum dose of 20 mg is administered.

Clinical Use

The great advantage of adenosine is its rapid onset of action, as well as the drug's brief duration of action, enabling rapid termination of effect. While it may cause ventricular standstill in patients with atrial flutter, the effect is momentary, to the physician's and patient's relief.

Digoxin

The digitalis glycosides have been around for over 200 years and are some of the oldest cardiovascular agents. Digoxin is one of the most frequently prescribed medications in the treatment of CHF and atrial arrhythmias.

Pharmacology/Electrophysiology

Digoxin exerts its cardiovascular effects in part through cholinergic stimulation and directly on cardiac cells (224). Digoxin's principal electrophysiologic effect is slowing of the AV node conduction and prolongation of the AV node refractory period. Sinus node automaticity is also slowed, mediated principally through the parasympathetic nervous system. The ECG effects are seen as PR prolongation and ST and T wave changes—the "dig effect."

Pharmacokinetics

The absorption of digoxin varies depending on the formulation (225). The bioavailability for the tablets is 60% to 80%; for capsules, 70% to 80%; and for the elixir, 90% to 100%. The VD of digoxin is 71/kg and follows a two-compartment distribution model. Approximately 20% to 25% of digoxin is protein bound. Partial breakdown of digoxin occurs in the gut in approximately 10% of patients due to bacterial action. Digoxin is cleared principally by the kidneys. The elimination half-life of digoxin is 20 to 30 hours in healthy volunteers. In patients with renal dysfunction as well as when patients age, the elimination half-life increases, at times requiring a dosage adjustment. Therapeutic serum concentration of digoxin ranges from 0.9 to 2.2 ng/ml. Patients with serum concentration >2 ng/ml can present with signs of toxicity, although many may be asymptomatic, and the concept of toxicity based solely on drug concentration may not be appropriate.

Clinical Use

Digoxin is primarily used in the treatment of atrial fibrillation and paroxysmal supraventricular tachycardia. Digoxin has been demonstrated to be as effective as the beta-blockers or calcium channel blocking agents in the treatment of these arrhythmias, although the agent does not convert AF to NSR (202). Digoxin is ineffective in the treatment of Wolff-Parkinson-White syndrome since it primarily alters antegrade conduction at the AV node. Digoxin, by shortening atrial refractoriness and blocking AV node, may accelerate the ventricular response, leading to hemodynamic collapse. Patients who are undergoing concurrent electrical cardioversion while receiving digoxin may be placed at higher risk for ventricular arrhythmias, though the incremental risk is small. Patients who present with elevated serum digoxin concentrations are at higher risk for developing ventricular arrhythmias [delayed afterdepolarizations (DADs), triggering ventricular tachycardia].

Adverse Effects

Acute adverse effects that are often seen with digoxin are early signs of toxicity. Adverse effects include a wide variety of cardiac and noncardiac manifestations. Cardiac effects include ventricular premature beats, atrial premature beats, ventricular tachycardia, ventricular bigeminy and trigeminy, junctional rhythm, and second- or third-degree AV block. Noncardiac effects include nausea and vomiting, anorexia, malaise, fatigue, delirium, and seizures.

Digoxin toxicity is less frequent than it once was but still can be a problem, especially in the elderly. Treatment of toxicity is supportive; activated charcoal may be employed for acute overdose. Arrhythmias commonly encountered include ventricular tachycardia, ventricular fibrillation, and complete AV block with slow ventricular escape rhythm and asystole. Lidocaine and phenytoin are used to treat the arrhythmias. Other agents used include beta-blockers, magnesium sulfate, procainamide, and atropine. Cardiac pacing for bradyarrhythmias may be necessary. Rapid and successful reversal of acute toxicity can be accomplished with the use of digoxin-specific antibodies (Fab fragments) (226,227). The use of the antibody is indicated in patients with VT, junctural tachycardias, and high-degree AV block. Also, those who have injected an overdose with high serum digoxin levels in the potentially life-threatening range may benefit from early antibody use.

While hypokalemia and hypomagnesia may facilitate the development of digitalis

toxicity, caution is required in the digitalis toxic patient. Potassium that is replaced may exacerbate acute high-degree AV blocks. The excess of digitalis inhibits Na-K—-adenosine trophosphatase (ATPase) and thus increases extracellular K^+ concentration. Replacing K^+ in these patients may lead to serious hyperkalemia and asystole.

Digoxin remains unique in that it is the only drug that prolongs refractoriness at the AV node while augmentating myocardial contractility. This unique combination of action makes digoxin a valuable therapy for the purpose of rate control in atrial flutter and fibrillation. However, the role of digoxin in long-term therapy for CHF is more circumspect. Digoxin as CHF therapy has been controversial for over a century. The legendary British cardiologist Sir John Mackenzie refused to use digitalis except in the context of atrial fibrillation. The recent Prospective Randomized Study of Ventricular Failure and the Efficacy of Dioxin (or PROVED) and Randomized Assessment of the Effort of Dioxin on Inhibiting the Angiotenism Converting Enzyme Study or RADI-ANCE trials suggest a benefit in CHF therapy for digoxin both with and without concomitant ACE inhibitor therapy (228,229). The DIG trial randomized 6,800 patients with heart failure in sinus rhythm who had symptoms of heart failure and left ventricular EF <0.45 with an average follow-up of 37 months (230). There was no significant difference between the two groups with regard to all-cause mortality. There was statistically significant reduction in death due to heart failure and acute hospitalizations. The data suggest that this benefit in CHF reduction was offset by a modest but clinically significant proarrhythmic action of digoxin canceling out the benefit in terms of total mortality.

Dosage and Administration

Digoxin is available both orally and intravenously. In the CCU setting the use of the IV formulation may be initially used for rapid digitalization, though this is rarely indicated currently. The suggested dose is 12 to 15 μg/kg over 24 hours administered in three equally divided doses. This is then followed by oral maintenance dose of 0.125 to 0.25 mg/daily. In patients with difficult-to-control atrial fibrillation and with the exclusion of thyrotoxicosis as an etiology, patients may receive up to 0.5 mg/day with monitoring of serum digoxin levels. When treating atrial flutter or fibrillation, the ventricular response is often adequate to gauge digoxin administration. Still one must be cautious to not overdigitalize. It is still possible to make the patient digitalis toxic when the patient is in AF. In sinus rhythm no guide exists to establish adequate digitalization. In sinus rhythm measurement of the serum digoxin levels, at least 4 hours from the last digoxin dose is effective in guiding therapy and avoiding toxicity.

Magnesium

Magnesium has been known to possess antiarrhythmic properties (231). Its use as an antiarrhythmic with its success in reducing sudden death in patients following a myocardial infarction has been reported (232,233). Acute administration in cardiac arrest patients reduced the incidence of arrhythmias (234).

Pharmacology/Electrophysiology

Magnesium is an intracellular cation that is distributed in both muscle and bone. Magnesium is involved in variety of physiologic actions providing maintenance of electrical homeostasis (235). Alteration in the magnesium serum concentration has been demonstrated to affect the electrical field strength and membrane excitability. Magnesium slows the conduction through the AV node and prolongs the refractory period in the atria and ventricles (231). Magnesium does not alter the sinus node function or ventricular refractory period.

Pharmacokinetics

Magnesium may be administered orally, intramuscularly, or intravenously. Intravenous administration achieves the fastest and most rapid onset of action. Magnesium is principally eliminated via the kidneys (235). Due to significant reabsorption in the kidney only 2% to 3% of magnesium is eliminated in the urine.

Clinical Use

The administration of magnesium to patients has been reported effective in supraventricular (236) as well as ventricular arrhythmias (234). There are many case reports that magnesium may be effective in patients with life-threatening ventricular tachycardia unresponsive to other antiarrhythmic agents. Magnesium has variable effects on APD (237). Magnesium administration has been found effective in reducing sudden death post-MI (233) in patients with polymorphic VT, in patients with the long QT syndrome (238), and in cases of digitalis toxicity (239). While magnesium would not be considered a first-line therapy or a chronic therapy, the acute administration of magnesium in patients unresponsive to other therapeutic modalities may be beneficial.

The International Study of Infarct Survival (ISIS) had an arm that administered magnesium in an open-label fashion (240). This study found that Mg^{2+} offered no advantage in terms of mortality outcome and did not reduce arrhythmic death. However, this was an open-label study permitting investigator bias. A review of magnesium as an antiarrhythmic has been compiled by Antman et al. (241,242).

Adverse Effects

The major complication associated with magnesium and the most frequent adverse effect following bolus or continuous infusion is hypotension. Often this can be rapidly reversed by decreasing the infusion rate. Hypotension in patients minimally compensated may lead to hemodynamic collapse, ischemia, ventricular fibrillation, and death.

Aggressive treatment of the hypotension is indicated if hemodynamic collapse develops. Other adverse effects include myocardial depression, hypothermia, and coma.

Dosage and Administration

Both the sulfate and chloride salt formulation of magnesium have been used successfully. The total dose of magnesium has varied considerably. In general, doses of 1 to 2 g are administered initially followed by a continuous infusion of 0.5 to 1 g/hour.

REFERENCES

1. Vaughan Williams EM. A classification of antiarrhythmic actions reassessed after a decade of new drugs. J Clin Pharmacol 1984;24:129.
2. Natel S. Antiarrhythmic drug classifications. A critical appraisal of their history, present status and clinical reference. Drugs 1991;41(5):672.
3. Ling GN, Gerald RW. The normal membrane potential of frog sartorius fibers. J Cell Comp Physiol 1949;34:383.
4. Willus FA, Keys TE. Cardiac clinics XCIV: a remarkable easy reference to the use of cinchona in cardiac arrhythmia. Mayo Clinic Proc 1942;17:249.
5. Mason JW, Hondeghem LM. Quinidine. Ann NY Acad Sci 1984;432:162.
6. Mason JW, Winkle RA, Rider AK, et al. The electrophysiologic effects of quinidine in the transplanted human heart. J Clin Invest 1977;59:481.
7. Mirro MJ, Watanabe AM, and Baily JC. Electrophysiologic effects of the optical isomer of disopyramide and quinidine in the dog: dependence on stereochemistry. Circ Res 1981;48:867.
8. Torres V, Flowers D, Miura D, et al. Intravenous quinidine by intermittent bolus for electrophysiologic studies in patients with ventricular tachycardia. Am Heart J 1984;108:1437.
9. Ueda CT, Hirschfeld DS, Scheinman MM, et al. Disposition kinetics of quinidine. Clin Pharmacol Ther 1976;19:30.
10. Fremstad D, Bergerud K, Haffner JF, et al. Increased plasma binding of quinidine after surgery: a preliminary report. Eur J Clin Pharmacol 1976;10:441.
11. Kessler KM, Lisker B, Conde C, et al. Abnormal quinidine binding in survivors of prehospital cardiac arrest. Am Heart J 1984;107:665.
12. Garfinkel D, Mameluk RD, Blascke TF. Altered therapeutic range for quinidine after myocardial infarction and cardiac surgery. Ann Intern Med 1987;107:48.
13. Kessler KM, Humphries WC, Black M, et al. Quinidine pharmacokinetics in patients with cirrhosis or receiving propranolol. Am Heart J 1978;96:627.
14. Ruskin JN, Dimarco JP, Garan H. Out of hospital cardiac arrest: electrophysiological observations and selection of long term therapy. N Engl J Med 1980;303:607.
15. Rossi M, Lown B. The use of quinidine in cardio version. Am J Med 1967;19:234.
16. Reynolds EW, Vander Ark CR. Quinidine syncope and the delayed repolarization syndromes. Mod Concepts Cardiovasc Dis 1976;45:117.
17. Hii JTY, Wyse DG, Gillis AM, Duff HJ, Solylo MA, Mitchel LB. Precordial QT interval dispersion marker of torsades de pointes. Circulation 1992;86:1376.
18. Hardy B, Zador IT, Golden L, et al. Effect of cimetidine on the pharmacodynamics of quinidine. Am J Cardiol 1983;52:172.
19. Twum-Barima Y, Carruthers SG. Quinidine-rifampin interaction. N Engl J Med 1981;304:1466.
20. Leahe EB Jr, Reiffel JA, Drusin RE, et al. Interaction between digoxin and quinidine. JAMA 1978;240:533.
21. Somberg JC, Knox S, Miura DS. The effect of quinidine on the differing sensitivities of Purkinje fibers and myocardium to inhibition of monovalent cation transport by digitalis. Am J Cardiol 1983;52:123.
22. Swerdlow CD, Yu Jo, Jacobson E, et al. Safety and efficacy of intravenous quinidine. Am J Med 1983;75:36.
23. Coplen SE, Antman EM, Berlin JA, Hewitt P, Chalmers TC. Efficacy and safety of quinidine therapy for maintenance of sinus rhythm after cardioversion. Circulation 1990;82:1106.
24. Stroke Prevention in Atrial Fibrillation Investigators. Stroke Prevention in Atrial Fibrillation Study; final results. Circulation 1991;84:527.
25. Josephson ME, Caracta AR, Ricciotti MA, et al. Electrophysiologic properties of procainamide in man. Am J Cardiol 1974;33:596.
26. Reidenberg MM, Drayer DE, Levy M. Polymorphic acetylation of procainamide in man. Clin Pharmacol Ther 1975;17:722.
27. Woosley RL, Drayer DE, Reidenberg MM, et al. Effect of acetylation phenotype on the rate at which procainamide induces antinuclear antibodies and lupus syndrome. N Engl J Med 1978;298:1157.
28. Wu D, Denes T, Bauernfiend R, et al. Effects of procainamide on atrial ventricular nodal re-entrant paroxysmal tachycardia. Circulation 1978;57:1171.
29. Strasbert B. Sclarovsky S, Erdberg A, et al. Procainamide induced polymorphous ventricular tachycardia. Am J Cardiol 1981;47:1309.
30. Josephson ME, Caracta AR, Lau SH et al. Electrophysiological evaluation of disopyramide in man. Am Heart J 1973;86:771.
31. LaBarre A, Strauss HC, Scheiman MM, et al. Electrophysiologic effects of disopyramide phosphate on sinus node function in patients with sinus node dysfunction. Circulation 1979;59:226–235.
32. Kates RE. Metabolites of cardiac antiarrhythmic drugs: their clinical roles. Ann NY Acad Sci 1984;432:75.
33. Bredesen JE, Kierulf P. Relationship between α-1-acid glycoprotein and plasma binding of disopyramide and mono-N-dealkyldiospyramide. Br J Clin Pharmacol 1984;18:779.
34. Haughey DB, Lima JJ. Influence of concentration dependent protein binding on serum concentrations and urinary excretion of disopyramide and its metabolite following oral administration. Biopharm Drug Dispos 1983;4:103.

35. Josephson ME, Horowitz LN. Electrophysiologic approach to therapy of recurrent sustained ventricular tachycardia. *Am J Cardiol* 1979;43:631.

36. Zainal N, Carmichael DJS, Griffiths JW, et al. Oral disopyramide for the prevention of arrhythmias in patients with acute myocardial infarction admitted to open wards. *Lancet* 1977;2:887.

37. Hartel G, Louhija A, Konttinen A. Disopyramide in the prevention of recurrence of atrial fibrillation after electroconversion. *Clin Pharmacol Ther* 1974;15:551.

38. Spurrell RAJ, Thorburn CW, Camm J, et al. Effects of disopyramide on electrophysiologic properties of specialized conduction system in man and on accessory atrio-ventricular pathway in the Wolff-Parkinson-White syndrome. *Br Heart J* 1975;37:861.

39. Podrid PG, Schoenebeyer A, Lown B. Congestive heart failure caused by oral disopyramide. *N Engl J Med* 1980;302:614.

40. Riccioni N, Castiglioni M, Bartolomei C. Disopyramide induced QT prolongation and ventricular tachyarrhythmias. *Am Heart J* 1983;105:870.

41. Davis LD, Temte JV. Electrophysiologic actions of lidocaine on canine ventricular muscle and Purkinje fibers. *Circ Res* 1969;24:639.

42. Josephson ME, Caracta AR, Lau SH, et al. Effects of lidocaine on refractory period of man. *Am Heart J* 1972;84:778.

43. Knapp AB, Maguire W, Keren, et al. The cimetidine-lidocaine interaction. *Ann Intern Med* 1983;98(2):174.

44. Haynes R, Chinn T, Copass M, Cobb L. Comparison of bretylium tosylate and lidocaine in management of out-of-hospital ventricular fibrillation: a randomized clinical trial. *Am J Cardiol* 1981;48:353.

45. Lie K, Wellen H, VanCapelle F, Durrer D. Lidocaine in the prevention of primary ventricular fibrillation. A double blind randomized study of 212 consecutive patients. *N Engl J* Med 1974;291:1324.

46. MacMahon S, Collins R, Peto R, et al. Effects of prophylactic lidocaine in suspected acute myocardial infarction. *JAMA* 1988;260:1910.

47. Wyse EDG, Kellen J, Rademaker A. Prophylactic versus selective lidocaine for early ventricular arrhythmias of myocardial infarction. *J Am Col Cardiol* 1988;12:507.

48. Yusuf S, Sleight P, Held P, et al. Routine medical management of myocardial infarction. *Circulation* 1990;82:117.

49. Nasir N Jr., Taylor A, Doyle TK and Pacifico A. Evaluation of intravenous lidocaine for the termination of sustained monomorphic ventricular tachycardia in patients with coronary artery disease with or without healed myocardial infarction. *Am J Cardiol* 1994;74:1183.

50. Akhtan M, Gilbert CJ, Shenasa M. Effect of lidocaine on atrioventricular response via the accessory pathway in patients with Wolff-Parkinson-White syndrome. *Circulation* 1981;63:435.

51. McComish M, Robinson C, Kitson D, et al. Clinical electrophysiological effects of mexilitene. *Postgrad Med* J 1977;53:85.

52. Roos JC, Paalman DCA, Dunning AJ. Electrophysiological effects of mexiletine in man. *Postgrad Med J* 1977;53:92.

53. Campbell NPS, Kell JG, Adgey AAJ, et al. The clinical pharmacology of mexiletine. *Br J Clin Pharmacol* 1978;6:103.

54. Monk JP, Brogden RN. Mexilitene. A review of its pharmacodynamic and pharmacokinetic properties, and therapeutic use in the treatment of arrhythmias. *Drugs* 1990;40:374.

55. Talbot RG, Julian DG, Prescott LF. Long term treatment of ventricular arrhythmias with oral mexiletine. *Am Heart J* 1976;91:58.

56. Lange R, Lee T, Wong K, et al. Mexiletine in the treatments of recurrent ventricular tachycardia. Prediction of long term arrhythmia suppression from acute and short term response. *J Clin Pharmacol* 1983;23:89.

57. DiMarco JP, Garan H, Ruskin GN. Mexiletine for refractory ventricular arrhythmias: results using serial electrophysiologic testing. *Am J Cardiol* 1981;47:131.

58. Duff HJ, Rosen D, Drimm K, et al. Mexiletine in the treatment of resistant ventricular arrhythmias, enhancement of efficacy in reduction of dose related side effects in combination with quinidine. *Circulation* 1983;67:1124.

59. Anderson JL, Mason JW, Winkle RA, et al. Clinical electrophysiology of tocainide. *Circulation* 1978;57:685.

60. Horowitz LN, Josephson ME, Farshidi A. Human electrophysiology of tocainide, a lidocaine congener. *Am J Cardiol* 1978;42:276.

61. Lalka D, Meyer M, Duce B, et al. Kinetics of the oral antiarrhythmic, lidocaine, congener tocainide. *Clin Pharmacol Ther* 1976;19:757.

62. Woosley RL, McDermott DG, Nies AS, et al. Suppression of ventricular ectopic depolarization by tocainide. *Circulation* 1977;56:980.

63. Winkle R, Mason JW, Harrison DC. Tocainide for drug resistant ventricular arrhythmias: efficacy, side effects and lidocaine responsiveness for predicting tocainide success. *Am Heart J* 1980;100:1041.

64. Maloney JD, Nissen RG, McColgan GM. Open clinical studies at a referral center: chronic maintenance tocainide therapy in patients with current sustained ventricular tachycardia refractory to conventional antiarrhythmic agents. *Am Heart J* 1980;100:1023.

65. Ikram H. Hemodynamic and electrophysiologic interaction between antiarrhythmic drugs and beta blockers with special reference to tocainide. *Am Heart J* 1980;100:1076.

66. Haffajee CI, Alpert JS, Dalen GE. Tocainide for refractory ventricular arrhythmias of myocardial infarction. *Am Heart J* 1980;100:1013.

67. Bastian BC, Macfarlane PW, McLaughlan JH, et al. A prospective randomized trial of tocainide in patients following myocardial infarction. *Am Heart J* 1980;100:1017.

68. Ryden L, Arnman K, Conradson T, et al. Prophylaxis of ventricular tachyarrhythmias with intravenous and oral tocainide in patients with and recovering from acute myocardial infarction. *Am Heart J* 1980;100:1006.

69. Morganroth J, Harlen S, MacVargh H, et al. Lidocaine in the treatment of ventricular arrhythmias after open heart surgery. *J Am Col Cardiol* 1983;1:700(A).

70. Horn H, Haddian Z, Johnson J, et al. Safety evaluation of tocainide in the American emergency use program. *Am Heart J* 1980;100:1037.

71. Jackman WM, Zipes DP, Naccarelli GU, et al. Electrophysiology of oral encainide. *Am J Cardiol* 1982;49:1270.

72. Winkle RA, Peters F, Kates RE, et al. Possible contribution of encainide metabolites to the long term antiarrhythmic efficacy of encainide. *Am J Cardiol* 1983;51:1182.

73. Winkle R, Peters F, Kates R, et al. Clinical pharmacology and antiarrhythmic efficacy of encainide in patients with chronic ventricular arrhythmias. *Circulation,* 1981;64:290.

74. Roden D, Reele S, Higgens S, et al. Total suppression of ventricular arrhythmias by encainide. *N Engl J Med* 1980;302:877.

75. The Cardiac Arrhythmia Suppression Trial (CAST) Investigators. Effect of encainide and flecainide on mortality in a randomized trial of arrhythmia suppression after myocardial infarction. *N Engl J Med* 1989;321:406.

76. Soyka LF. Safety of encainide for the treatment of ventricular arrhythmias. *Am J Cardiol* 1986;58:96C.

77. Hellestrand KJ, Bexton RS, Nathan AW, et al. Acute electrophysiological effects of flecainide acetate on conduction and refractoriness in man. *Br Heart J* 1982;48:140.

78. Conrad GJ, Ober RE. Metabolism of flecainide. *Am J Cardiol* 1984;53:41B.

79. Franciosa JA, Wilen M, Weeks CE, et al. Pharmacokinetics and hemodynamic effects of flecainide in patients with chronic low output heart failure. *J Am Col Cardiol* 1983;1:669A.

80. Banitt EF, Bronn WR, Coyne WE, et al. Antiarrhythmia synthesis and antiarrhythmic N-(piperidylalkyl) trifluroethoxybenzamides. *J Med Chem* 1977;20:821.

81. Anderson JL, Stewart JR, Perry BA, et al. Oral flecainide acetates for the treatment of ventricular arrhythmias. *N Engl J Med* 1981;305:473.

82. Flowers D, O'Gallagher D, Torres V, et al. Flecainide: long term treatment using a reduced dosing schedule. *Am J Cardiol* 1985;55:79.

83. Hellestrand KJ., Nathan AW, Bexton RS, et al. Electrophysiologic effects of flecainide acetate on sinus node function, anomalous atrioventricular connections and pacemaker thresholds. *Am J Cardiol* 1984;53:30B.

84. Hellestrand KJ. Intravenous flecainide acetate for supraventricular tachycardias. *Am J Cardiol* 1988;62:16D.

85. Anderson JL, Platt ML, Guarnieri T, Fox TL, Maser MJ, Pritchett ELC, and the Flecainide Supraventricular Tachycardia Study Group. Flecainide acetate for paroxysmal supraventricular tachyarrhythmias. *Am J Cardiol* 1994;74:578.

86. Henthorn RW, Waldo AL, Anderson JL, Gilbert EM, Alpert BL, Bhandari AK, Hawkinson RW, Pritchett ELC, and the Flecainide Supraventricular Tachycardia Group. Flecainide acetate prevents recurrence of symptomatic paroxysmal supraventricular tachycardia. *Circulation* 1991;83:119.

87. Pritchett EL, DaTorre SD, Platt ML, McCarville SE, Hougham AJ. Flecainide acetate treatment of paroxysmal supraventricular tachycardia and paroxysmal atrial fibrillation: dose-response studies. *J Am Coll Cardiol* 1991;17:297.

88. Bhandari AK, Anderson JL, Gilbert EM, Alpert BL, Henthorn RW, et al. Correlation of symptoms with occurrence of paroxysmal supraventricular tachycardia or atrial fibrillation: a transtelephonic monitoring study. *Am Heart J* 1992;124:381.

89. Schwartz J, Crocker K, Somberg J. Refractoriness as a determinant of right ventricular inducibility. *Clin Res* 1987;35:325A.

90. Ranger S, Talajic M, Lemery R, Roy D, Nattel S. Amplification of flecainide-induced ventricular conduction slowing by exercise. *Circulation* 1989;79:1000.

91. Ledda I, Mantelli L, Manzini S, et al. Electrophysiologic and antiarrhythmic properties of propafenone in isolated cardiac preparations. *J Cardiovasc Pharmacol* 1981;3:1162.

92. Karaqueuzian HS, Kato T, Sugi K, et al. Electrophysiologic effects of propafenone, a new antiarrhythmic drug, on isolated cardiac tissue. *Circulation* 1982;66:375(A).

93. Breithardt G, Burggrefe M, Wiebringhaus E, et al. Effect of propafenone in the Wolfe Parkinson White syndrome. Electrophysiologic findings and long term follow-up. *Am J Cardiol* 1984;54:29D.

94. Baker BJ, Desoyza N, Boyd CM, et al. Effects of propafenone on left ventricular function (abstract). *Circulation* 1982;67:II-267.

95. Somberg JC, Tepper D, Landau S. Propafenone: a new antiarrhythmic agent. *Am Heart J* 1988;115:1274.

96. Siddoway LA, McAllister CB, Want T, et al. Polymorphic oxidative metabolism of propafenone in man. *Circulation* 1983;68:64(A).

97. Frank R, Tonet JL, Lacroix H, et al. Electrophysiological effects and efficacy of oral propafenone in the Wolff Parkinson White syndrome. *Circulation* 1984;70:442(A).

98. Shen EN, Keung E, Huyeke E, et al. Intravenous propafenone for the termination of re-entrant supraventricular tachycardia: a placebo-controlled, randomized, double blind, crossover study. *Ann Intern Med* 1986;105:655.

99. Bianconi L, Mennuni M, Lukic V, Castro A, Chieffi M, Santini M. Effects of oral propafenone administration before electrical cardioversion of chronic atrial fibrillation: a placebo-controlled study. *J Am Coll Cardiol* 1996;28:700.

100. Doherty JU, Waxman HL, Kienzle MG, et al. Limited role of intravenous propafenone hydrochloride in the treatment of sustained ventricular tachycardia: electrophysiological effects and results of programmed ventricular stimulation. *J Am Col Cardiol* 1984;4:378.

101. Siebels J, Kuch KH. Implantable cardioverter dexibrillator compared with antiarrhythmic drug treatment in cardiac arrest survivors (The Cardiac Arrest Study, Hamburg) *American Heart Journal* 1994;127:1139–1144.

102. Vaughan Williams EM. Classification of the antiarrhythmic action of moricizine. *J Clin Pharmacol* 1991;31:216.

103. Wyndham CRC, Pratt CM, Mann D, et al. Electrophysiology of ethmozine (moricizine HCL) for ventricular tachycardia. *Am J Cardiol* 1987;60:67F.

104. Woosely RL, Morganroth J, Fogoros RN, et al. Pharmacokinetics of moricizine. *Am J Cardiol* 1987;35F.

105. Siddoway LA, Schwartz SL, Barbey JT, et al. Clinical pharmacokinetics of moricizine. *Am J Cardiol* 1990;65:21D.

106. Howrie DL, Pieniaszek HJ, Fogoros RN, et al. Disposition of moricizine (ethmozine) in healthy subjects

after oral administration of radiolabelled drug. *Eur J Clin Pharmacol* 1987;32:607.

107. Podrid PJ, Lyakisheu A, Lown B, et al. Ethmozine, a new antiarrhythmic drug for suppressing ventricular premature complexes. *Circulation* 1980;61:450.

108. Pratt CM, Yepsen SC, Taylor AA, et al. Ethmozine suppression of single and repetitive ventricular premature depolarization during therapy: documentation of efficacy and long-term safety. *Am Heart J* 1983; 106:85.

109. Morganroth J, Pratt CM, Kennedy HL, et al. Efficacy and tolerance of ethmozine (moricizine HCL) in placebo controlled trials. *Am J Cardiol* 1987;60:48F.

110. Cardiac Arrhythmias Suppression Trial (CAST) Investigators. Effect of the antiarrhythmic agent moricizine on survival after myocardial infarction. *N Engl J Med* 1992;327:227.

111. Miura D, Wynn J, Torres V, et al. Antiarrhythmic efficacy of ethmozine in patients with ventricular tachycardia as determined by programmed electrical stimulation. *Am Heart J* 1985;111:661.

112. Molinoff PB. Evolving properties of β-adrenergic receptor antagonists. *Pharmacotherapy* 1992;12:144.

113. Upward JW, Waller DG, George CF. Class II antiarrhythmic agents. *Pharmacol Ther* 1988;37:81.

114. Gray RJ, Bateman TM, Czer LSC, et al. Esmolol: a new ultrashort-acting beta-adrenergic blocking agent for rapid control of heart rate in postoperative supraventricular tachyarrhythmias. *J Am Col Cardiol* 1985;5: 1451.

115. Olley PM, Fowler RS. The pseudo-cardiac syndrome and therapeutic observations. *Br Heart J* 1970;32:467.

116. Reiffel J. Improved rate control in atrial fibrillation: editorial. *Am Heart J* 1992;123:1094.

117. Sloskey GE. Amiodarone: a unique antiarrhythmic agent. *Clin Pharmacol* 1983;2:330.

118. Torres V, Tepper D, Flowers D, Wynn J, et al. QT prolongation and the antiarrhythmic efficacy of amiodarone. *J Am Col Cardiol* 1986;7:142.

119. Riva E, Gerna M, Latini R, et al. Pharmacokinetics of amiodarone in man. *J Cardiovasc Pharmacol* 1982;4: 264.

120. Puech P. Practical aspects of the use of amiodarone. *Drugs* 1991;41:67.

121. Mostow N, Rakita L, Blumer J. Amiodarone: correlation of serum concentration with clinical efficacy. *Circulation* 1982;66:223A.

122. Rosenbaum MB, Chiale PA, Halpern MS. Clinical efficacy of amiodarone as an antiarrhythmic agent. *Am J Cardiol* 1976;38:934.

123. Kaski JC, Girotti LA, Mesuti H. Long-term management of sustained, recurrent, symptomatic ventricular tachycardia with amiodarone. *Circulation* 1981;64: 273.

124. Nademanee K, Hendrixing JA, Cannom DS, et al. Control of refractory life threatening ventricular tachycardias by amiodarone. *Am Heart J* 1981;101:759.

125. Ward DE, Camm AJ, Spurrell RA. Clinical antiarrhythmic effects of amiodarone in patients with resistant paroxysmal tachycardia. *Br Heart J* 1980;44:91.

126. Fogoros RN, Anderson KP, Winkle RA, et al. Amiodarone, clinical efficacy and toxicity in 96 patients with recurrent drug refractory arrhythmias. *Circulation* 1983;68:88.

127. Chun, SH, Sager PT, Stevenson WG, Nademanee K, Middlekauff HR, Singh BN. Long-term efficacy of amiodarone for the maintenance of normal sinus rhythm in patients with refractory atrial fibrillation or flutter. *Am J Cardiol* 1995;76:47.

128. Zarembski D, Nolan PE Jr., Slack MK, Caruso AC. Treatment of resistant atrial fibrillation; a meta-analysis comparing amiodarone and flecainide. *Arch Intern Med* 1995;155:1885.

129. Galve E, Rius T, Ballester R, Artaza MA, Arnau JM, Garcia-Dorado D, Soler-Soler J. Intravenous amiodarone in treatment of recent-onset atrial fibrillation: results of a randomized, controlled study. *J Am Coll Cardiol* 1996;27:1079.

130. Horowitz LN, Greenspan AM, Spielman SR, et al. Usefulness of electrophysiologic testing in evaluation of amiodarone therapy for sustained ventricular tachyarrhythmias associated with coronary heart disease. *Am J Cardiol* 1985;55:367.

131. Gottlieb CD, Slivka T, Langan MN, et al. Do implantable defibrillators prolong survival compared to electrophysiologic guided therapy. *Circulation* 1982;86:655A.

132. CASCADE Investigators. Randomized antiarrhythmic drug therapy in survivors of cardiac arrest (the CASCADE study). *Am J Cardiol* 1993;72:280.

133. Burkart F, Pfisterer M, Kiowski W, Follath F, Burckhardt D. Effect of antiarrhythmic therapy on mortality in survivors of myocardial infarction with asymptomatic complex ventricular arrhythmias: Basal Antiarrhythmic Study of Infarct Survival (BASIS). *J Am Coll Cardiol* 1990;16:1711.

134. Ceremuzynski L. Secondary prevention after myocardial infarction with class III antiarrhythmic drugs. *Am J Cardiol* 1993;72:82F.

135. Camm AJ, Julian D, Janse G, Munoz A, Schwartz P, Simon P, Frangin G, on behalf of the EMIAT Investigators. The European Myocardial Infarct Amiodarone Trial (EMIAT). *Am J Cardiol* 1993;72:95F.

136. Camm AJ and the European Myocardial Infarct Amiodarone Trial (EMIAT). American Heart Association, New Orleans, LA, USA (oral presentation), November, 1996.

137. Cairns JA, Connolly SJ, Roberts R, Gent M, on behalf of the CAMIAT Investigators. Canadian Amiodarone Myocardial Infarction Arrhythmia Trial (CAMIAT): rationale and protocol. *Am J Cardiol* 1993;72;87F.

138. Van Hare GF, Waldo AL. The atrial flutter reentrant circuit; additional pieces of the puzzle. *Circulation* 1996;94:244.

139. Holmes J, Kubo SH, Cody RJ, Kligfield P. Arrhythmias in Ischemic and nonischemic dilated cardiomyopathy: prediction of mortality by ambulatory electrocardiography. *Am J Cardiol* 1985;55:146.

140. von Olshausen K, Schafer A, Mehmel HC, Schwarz F, Senges J, Kubler W. Ventricular arrhythmias in idiopathic dilated cardiomyopathy. *Br Heart J* 1984;51:195.

141. Cleland JGF, Dargie JH, Findlay IN, Wilson JT. Clinical, haemodynamic and antiarrhythmic effects of long term treatment with amiodarone of patients in heart failure. *Br Heart J* 1987;57:436.

142. Chatterjee K. Amiodarone in chronic heart failure: editorial. *J Am Coll Cardiol* 1989;14:1774.

143. Nicklas JM, McKenna WJ, Stewart RA, Mickelson JK, et al. Prospective, double-blind, placebo-controlled

trial of low-dose amiodarone in patients with severe heart failure and asymptomatic frequent ventricular extopy. *Am Heart J* 1991;122:1016.

144. Doval JC, Nul DR, Grancelli HO, Perrone SV, Bortman GR, Curiel R, GESICA. Randomized trial of low-dose amiodarone in severe congestive heart failure. *Lancet* 1994;344:493.

145. Singh SN, Fletcher RD, Fisher SG, et al. Amiodarone in patients with congestive heart failure and asymptomatic ventricular arrhythmia. *N Engl J Med* 1995;333:77.

146. Gorgels A, van den Dool A, Hofs A, Mulleneers R, Smeets J, Vos M, Wellens J. Comparison of procainamide and lidocaine in terminating sustained monomorphic ventricular tachycardia. *Am J Cardiol* 1996;78:43.

147. Munoz A, Karila P, Gallay P, Zettelmeier F, Messner P, Mery M, Grolleau R. A randomized hemodynamic comparison of intravenous amiodarone with and without Tween 80. *Eur Heart J* 1988;9:142.

148. Helmy I, Herre JM, Gee G, et al. Use of intravenous amiodarone for emergency treatment of life-threatening ventricular arrhythmias. *J Am Coll Cardiol* 1988; 12:1015.

149. Kowey PR, Marinchak RA, Rials SJ, Rubin AM, Smith L. Electrophysiologic testing in patients who respond acutely to intravenous amiodarone for incessant ventricular tachyarrhythmias. *Am Heart J* 1993;125:1628.

150. Levine JH, Massumi A, Scheinman MM et al. Intravenous amiodarone for recurrent sustained hypotensive ventricular tachyarrhythmias. *J Am Col Cardiol* 1996;27:67.

151. Kowey PR, Levine JH, Herre JM, et al. Randomized, Double-blind comparison of intravenous amiodarone and bretylium in the treatment of patients with recurrent, hemodynamically destabilizing ventricular tachycardia or fibrillation. *Circulation* 1995;92:3255.

152. Scheinman MH, Levine JH, Cannom DS, et al. Dose-ranging study of intravenous amiodarone in patients with life-threatening ventricular tachyarrhythmias. *Circulation* 1995;92:3264.

153. Somberg J, Cao W, Ranade V. A new aqueous preparation of amiodarone (abstract). *J Invest Med* 1995; 43(3):430A.

154. Somberg J, Cao W, Ranade V. A new aqueous preparation of amiodarone (abstract). *Circulation* 1995;92(8): I-195 (0933).

155. Gallik D, Meissner M, Singer I, Hanin L. Hemodynamic and electrocardiographic effects of amio-aqueous: a new formulation of intravenous amiodarone (abstract submitted to NASPE, 1997).

156. Mostow ND, Rakita L, Vrobel TR, Noon D, Blumer J. Amiodarone: intravenous loading for rapid suppression of complex ventricular arrhythmias. *J Am Coll Cardiol* 1984;4(1):97.

157. Blevins RD, Zerin NZ, Benaderet D, Frumin H, Faitel K, Jarandilla R, Rubenfire M. Amiodarone in the management of refractory atrial fibrillation. *Arch Intern Med* 1987;147:1401.

158. Horowitz LN, Spielman SR, Greenspan AM, et al. Use of amiodarone in the treatment of persistent and paroxysmal atrial fibrillation resistant to quinidine therapy. *J Am Coll* Cardiol 1985;6:1402.

159. Howard PA. Amiodarone for the maintenance of sinus rhythm in patients with atrial fibrillation. *Ann Pharmacother* 1995;29:596.

160. Molnar J, Jerry W, Rosenthal JE, Somberg JC. Circadian variation of QT dispersion in survivors of sudden cardiac death. *PACE* 1996;19(II):688A.

161. Marchlinski FE, Gansler TS, Waxman HL, et al. Amiodarone pulmonary toxicity. *Ann Intern Med* 1982;97:839.

162. Mayuga RD, Singer DH. Effects of intravenous amiodarone on electrical dispersion in normal and ischaemic tissues and on arrhythmia inducibility: monophasic action potential studies. *Cardiovasc Res* 1992;26:571.

163. Koch-Wesner J. Bretylium. *N Engl J Med* 1979;300: 473.

164. Rapaport WG. Clinical pharmacokinetics of bretylium. *Clin Pharmacokinet* 1985;10:248.

165. Anderson JL, Patterson E, Conlon M, et al. Kinetics and antifibrillatory effects of bretylium: correlation with myocardial drug concentrations. *Am J Cardiol* 1980;46:583.

166. Hayes RE, Chinn TL, Cupass MK, et al. Comparison of bretylium tosylate and lidocaine in management of out of hospital ventricular fibrillation. A randomized clinical trial. *Am J Cardiol* 1981;48:353.

167. Euler DE, Zeman TW, Wallock ME, et al. Deleterious effects of bretylium on hemodynamic recovery from ventricular fibrillation. *Am Heart J* 1986;112:25.

168. Woosely RL, Reele SB, Roden DM, et al. Pharmacological reversal of the hypotensive effect that complicates antiarrhythmic therapy with bretylium. *Clin Pharmacol Ther* 1982;32:313.

169. Touboul P, Atallah G, Kirkorian G, et al. Clinical electrophysiology of intravenous sotalol, a beta blocking drug, with class III antiarrhythmic properties. *Am Heart J* 1984;107(5):888.

170. Blair AD, Burgess ED, Maxwell BM, et al. Sotalol kinetics in renal insufficiency. *Clin Pharmacol Ther* 1981;29(4):457.

171. Rizos I, Senges J, Jauernig R, et al. Differential effects of sotalol and metoprolol on induction of paroxysmal supraventricular tachycardia. *Am J Cardiol* 1984;53: 1022.

172. Halinen MO, Huttunen M, Paakkinen S, Tarssanen L. Comparison of sotalol with digoxin-quinidine for conversion of acute atrial fibrillation to sinus rhythm (the Sotalol-Digoxin-Quinidine Trial). *Am J Cardiol* 1995; 76:495.

173. Reimold SC, Cantillon CO, Friedman PL, Antman EM. Propafenone versus sotalol for suppression of recurrent symptomatic atrial fibrillation. *Am J Cardiol* 1993;71:558.

174. Senges J, Lengfelder W, Javernig R, et al. Electrophysiologic testing of therapy with sotalol for sustained ventricular tachycardia. *Circulation* 1985;72:577.

175. Nademanee K, Feld G, Hendrickson JAJI et al. Electrophysiologic and antiarrhythmic effects of sotalol in patients with life-threatening ventricular tachyarrhythmias. *Circulation* 1985;72(3):555.

176. Nademanee K, Lee IK, Singh BN. Sotalol versus procainamide in the prevention of ventricular tachycardia induction: a double blind parallel multicenter study. *Circulation* 1986;74:1242A.

177. Mason J for ESVEM. A comparison of seven antiarrhythmic drugs in patients with ventricular tachyarrhythmias. *N Engl J Med* 1993;329:452.

178. Kato R, Ikeda N, Yabek S, Kannan R, Singh B. Electrophysiologic effects of the levo- and dextrorotatory isomers of sotalol in isolated cardiac muscle and their in vivo pharmacokinetics. *J Am Coll Cardiol* 1986;7:116.

179. Waldo AL, Camm AJ, deRuyter H, et al. Effect of d-sotalol on mortality in patients with left ventricular dysfunction after recent and remote myocardial infarction. *Lancet* 1996;348:7.

180. Julian DG, Jackson FS, Prescott RJ, Szekely P. Controlled trial of sotalol for one year after myocardial infarction. *Lancet* 1982;1:1142.

181. Soyka LF, Wirtz C, Spangenberg RB. Clinical safety profile of sotalol in patients with arrhythmias. *Am J Cardiol* 1990;65:74A.

182. Podrid PJ, Lampert S, Graboys TB, Blatt CM, Lown B. Aggravation of arrhythmia by antiarrhythmic drugs—incidence and predictors. *Am J Cardiol* 1987;59:38E.

183. Hohnloser SH, Klingenheben T, Singh BN. Amiodarone-associated proarrhythmic effects: a review with special reference to torsades de pointes tachycardia. *Ann Intern Med* 1994;121:529.

184. Lehmann MH, Hardy S, Archibald D, Quart B, MacNeil DJ. Sex difference in risk of torsades de pointes with d,l-sotalol. *Circulation* 1996;94:2534.

185. Makkar RR, Fromm BS, Steinman RT, Meissner MD, Lehmann MH. Female gender as a risk factor for torsades de pointes associated with cardiovascular drugs. *JAMA* 1993;270(21):2590.

186. Lee K. Ibutilide, a new compound with potent class III antiarrhythmic activity, activates a slow inward Na+ current in guinea pig ventricular cells. *J Pharmacol Exp Ther* 1992;262:99.

187. Yang T, Wathen MS, Felipe A, Tamkun MM, Snyders DJ, Roden DM. K+ currents and K+ channel mRNA in cultured atrial cardiac myocytes. *Circ Res* 1994;75;870.

188. Yang T, Snyders DJ, Roden DM. Ibutilide, a methanesulfonanilide antiarrhythmic, is a potent blocker of the rapidly activating delayed rectifier K+ current (Ikr) in AT-1 cells. *Circulation* 1995;91:1799.

189. Stambler BS, Wood MA, Ellenbogen KA, Perry KT, Wakefield LK, VanderLugt JT and the Ibuilide Repeat Dose Study Investigators. Efficacy and safety of repeated intravenous doses of ibuilide for rapid conversion of atrial flutter or fibrillation. *Circulation* 1996;94:1613.

190. Ellenbogen KA, Stambler BS, Wood MA, et al. Efficacy of intravenous ibutilide for rapid termination of atrial fibrillation and atrial flutter: a dose-response study. *J Am Coll Cardiol* 1996;28:130.

191. Bih-Fang G, Ellenbogen KA, Wood MA, Stambler BS. Conversion of atrial flutter of ibutilide is associated with increased atrial cycle length variability. *J Am Coll Cardiol* 1996;27:1083.

192. Mitchell LB, Schroeder JS, Mason JW. Comparative clinical electrophysiologic effects of diltiazem, verapamil and nifedipine: a review. *Am J Cardiol* 1982;49:629.

193. Kates RE. Calcium antagonists: pharmacokinetic properties. *Drugs* 1985;28:405.

194. Schamroth L, Krikler DM, Garrett C. Immediate effects of intravenous verapamil in cardiac arrhythmias. *Br Heart J* 1972;1:660.

195. Schamroth L. Immediate effects of intravenous verapamil on atrial fibrillation. *Cardiovasc Res* 1971;5:419.

196. Waxman HL, Meyerberg RJ, Apple R, et al. Verapamil for control of ventricular rate in paroxysmal supraventricular tachycardia and atrial fibrillation or flutter. *Ann Intern Med* 1981;94:1.

197. Heng MK, Singh BN, Roche AH, et al. Effects of intravenous verapamil on cardiac arrhythmias and on the electrocardiogram. *Am Heart J* 1975;90:487.

198. Bender F, Reploh HD. Treatment of ventricular tachycardia with isoptin. *Med Klin* 1968;63:715.

199. Wellens HJ, Farre J, Bar FW. The role of the slow inward current in the genesis of ventricular tachyarrhythmias in man. In: Zipes, Bailey, Elharrar, eds. The slow inward current and cardiac arrhythmias. *Dev Cardiovasc Med* 1980;7:507.

200. Siegel L, Keren G, Tortes V, et al. Antiarrhythmic action of verapamil in preventing programmed stimulation induced ventricular tachycardia. *Clin Res* 1983;31:633A.

201. Falk, RH, Knowlton AA, Bernard SA, Gotlieb NE, Battinelli N. Digoxin for converting recent-onset atrial fibrillation to sinus rhythm. *Ann Intern Med* 1987;106:503.

202. Panidis IP, Morganroth J, Baessler C. Effectiveness and safety of oral verapamil to control exercise-induced tachycardia in patients with atrial fibrillation receiving digitalis. *Am J Cardiol* 1983;52:1197.

203. Weiss AT, Lewis BS, Halon DA, et al. The use of calcium with verapamil in the management of supraventricular tachyarrhythmias. *Int J Cardiol* 1983;4:275.

204. Gulamhusein S, Ko P, Carruthers G, Klein GJ. Acceleration of the ventricular response during atrial fibrillation in the Wolff-Parkinson-White syndrome after verapamil. *Circulation* 1982;65(2):348.

205. Psaty BM, Heckbert SR, Koepsell TD. The risk of myocardial infarction associated with antihypertensive drug therapies. *JAMA* 1995;274:620.

206. Nayler WG. Review: calcium antagonists and the ischemic myocardium. *Int J Cardiol* 1987;15:267.

207. Kloner R, Braunwald E. Effects of calcium antagonists on infarcting myocardium. *Am J Cardiol* 1987;59:84B.

208. Przyklenk K, Kloner RA. Effect of verapamil on postischemic stunned myocardium: importance of the timing of treatment. *J Am Coll Cardiol* 1988;11:614.

209. The Danish Study Group on Verapamil in Myocardial Infarction. Verapamil in acute myocardial infarction. *Eur Heart J* 1984;5:516.

210. The Danish Study Group on Verapamil in Myocardial Infarction. Effect of verapamil on mortality and major events after acute myocardial infarction (The Danish Verapamil Infarction Trial II—DAVIT II). *Am J Cardiol* 1990; 66:779.

211. Iberti TJ, Benjamin E, Paluch TA, et al. Use of constant-infusion verapamil for the treatment of postoperative supraventricular tachycardia. *Crit Care Med* 1986;14:283.

212. Kawai C, Konishi T, Matsuyama E, et al. Comparative effects of three calcium antagonists, diltiazem, verapamil, and nifedipine on the sinoatrial and atrioventricular nodes: experimental and clinical studies. *Circulation* 1981;63:1035.

213. Ellenbogen KA, Dias VC, Cardello FP, et al. Safety and efficacy of intravenous diltiazem in atrial fibrillation or atrial flutter. *Am J Cardiol* 1995;75:45.

214. Huycke E, Sung R, Dias V, et al. Intravenous diltiazem for termination of re-entrant supraventricular tachycar-

dia: a placebo-controlled, randomized double-blind multicenter study. *J Am Col Cardiol* 1989;13:538.

215. Salerna D, Dias V, Kleiger R, et al. Efficacy and safety of intravenous diltiazem for treatment of atrial fibrillation and atrial flutter. *Am J Cardiol* 1989;63:1046.

216. Ellenbogen KA, Dias VC, Plumb Vi, et al. A placebo-controlled trial of continuous intravenous diltiazem infusion for 24 hour heart rate control during atrial fibrillation and atrial flutter: a multicenter study. *J Am Col Cardiol* 1991;18:891.

217. Boden WE, Krone RJ, Kleiger RE, et al. Electrocardiographic subset analysis of diltiazem administration on long-term outcome after acute myocardial infarction. *Am J Cardiol* 1991;67:335.

218. Gibson RS, Boden WE, Theroux P, et al. Diltiazem and reinfarction in patients with non-Q-wave myocardial infarction. *N Engl J Med* 1986;315:423.

219. Faulds D, Crisp P, Buckley MT. Adenosine: an evaluation of its use in cardiac diagnostic procedures, and in the treatment of paroxysmal supraventricular tachycardia. *Drugs* 1991;41:596.

220. DiMarco JP, Sellers D, Berne RM, West GA, Belardinelli L. Adenosine: electrophysiologic effects and therapeutic use for terminating paroxysmal supraventricular tachycardia. *Circulation* 1983;68:(6) 1254.

221. Griffith MJ, Garratt CJ, Rowland E, Ward DE, Camm AJ. Effects of intravenous adenosine on verapamil-sensitive idiopathic ventricular tachycardia. *Am J Cardiol* 1994;73:759.

222. Bartlett TG, Friedman PL. Adenosine and Wolff-Parkinson-White. *J Cardiac Surg* 1993;8(4):503.

223. Sharma A, Klein GJ, Yee R. Intravenous adenosine triphosphate during wide QRS complex tachycardia: safety, therapeutic efficacy and diagnostic utility. *Am J Med* 1990;88:337.

224. Smith TW, Braunwald E, Kelley R. The management of heart failure. In: *Braunwald heart disease. A textbook of cardiovascular medicine,* 4th ed. Philadelphia: WB Saunders, 1992;464.

225. Smith TW. Pharmacokinetics, bioavailability and serum levels of cardiac glycosides. *J Am Col Cardiol* 1985;5:43A.

226. Wenger TL, Butler VP, Haber E, et al. Treatment of 63 severely digitalis-toxic patients with digoxin-specific antibody fragments. *J Am Col Cardiol* 1985;5:118A.

227. Ujheiyi MR, Colucci RD, Cummings DM, et al. Monitoring serum digoxin concentrations during digoxin immune lab therapy. *DICP Ann Pharmacother* 1991; 25:1047.

228. Uretsky BF, Young JB, Shahidi FE, Yellen LG, Harrison MC, Jolly MK. Randomized study assessing the effect of digoxin withdrawal in patients with mild to moderate chronic congestive heart failure: results of the PROVED trial. *J Am Coll Cardiol* 1993;22:955.

229. Packer M, Gheorghiade M, Young JB, et al. Withdrawal of digoxin from patients with chronic heart failure treated with angiotensin-converting-enzyme inhibitors. *N Engl J Med* 1993;329:1.

230. Grogan M, Smith HC, Gersh B, Wood DL. Left ventricular dysfunction due to atrial fibrillation in patients initially believed to have idiopathic dilated cardiomyopathy. *Am J Cardiol* 1992;69:1570.

231. Rasmussen HS, Thomsen PEB. The electrophysiological effects of intravenous magnesium on human sinus node, atrioventricular node, atrium, and ventricle. *Clin Cardiol* 1989;12:85.

232. Rasmussen HS, Norregard P, Lindeneg 0, et al. Intravenous magnesium in acute myocardial infarction. *Lancet* 1986;1:234.

233. Rasmussen HS, Gronbaet M, Cirrtin C, et al. One year death rate in 270 patients with suspected acute myocardial infarction initially treated with intravenous magnesium or placebo. *Clin Cardiol* 1988;11: 377.

234. Smith LF, Heagerty AM, Bing RF, et al. Intravenous infusion of magnesium sulfate after acute MI: effects on arrhythmias and mortality. *Int J Cardiol* 1986;12: 175.

235. Wacker WEC, Parisi AF. Magnesium metabolism. *N Engl J Med* 1968;278:658.

236. Wesley R, Haines D, Lerman B, et al. Effect of intravenous magnesium sulfate on supraventricular tachycardia. *Am J Cardiol* 1989;63:1129.

237. Matusda H, Saigusa A, Irisawa H. Ohmic conductance through the inward rectifying K channel and blocking by internal magnesium. *Nature* 1997;325:156.

238. Peticone F, Adinolfi L, Bonduce D. Efficacy of magnesium sulfate in the treatment of torsade de pointes with magnesium sulfate. *Circulation* 1988;77:392.

239. Kinlay S, Buckley NA. Magnesium sulfate in the treatment of ventricular arrhythmias due to digoxin toxicity. *J Toxicol Clin Toxicol* 1995;33(1):55.

240. ISIS-4 (Fourth International Study of Infarct Survival) Collaborative Group. ISIS-4: a randomised factorial trial assessing early oral captopril, oral mononitrate, and intravenous magnesium sulphate in 58050 patients with suspected acute myocardial infarction. *Lancet* 1995;345:669.

241. Antman EM. Magnesium in acute myocardial infarction: overview of available evidence. *Am Heart J* 1996; 132:487.

242. Antman EM. Magnesium in acute MI. Timing is critical. *Circulation* 1995;92:2367.

11

Clinical Therapy

John Somberg

An understanding of the basic electrophysiology and clinical pharmacology of antiarrhythmic drugs is most helpful for their clinical application. This knowledge combined with clinical experience aids the clinician in the selection of a therapy. Perhaps the most important facet is an understanding of the specific needs of a given patient and the underlying pathophysiology. The clinical data set is most important in determining the choice of an agent or possible alternative therapy.

The use of antiarrhythmic drugs has become limited for a number of reasons. The substrate of coronary artery disease is a changing one. Lipid reduction strategies, the introduction of the hepatic hydroxymethylglutaryl coenzyme A (HMG CoA) reductase inhibitors, and the use of aspirin combined with aggressive interventional cardiology including the use of thrombolytics and angioplasty have combined to reduce the prevalence of serious ventricular arrhythmias in the context of coronary artery disease. We have also learned more about the potential adversity of antiarrhythmic agents. The Cardiac Arrhythmia Suppression Trial (CAST) study has forever warned the prescribing physician of the possible mortal adversity that lies with antiarrhythmics. The Cardiac Arrhythmia Study—Hamburg and the Survival with Oral d-Sotalol (SWORD) studies have further reinforced this potential adversity with d-sotalol and propafenone. Along with these changes has been the ascendancy of the implantable defibrillator (ID).

The implantable defibrillator's development was pioneered by Michel Mirowski. The device has evolved markedly from its initial prototype. Initially, a large power supply and electrical component unit would be implanted in the abdomen and patches sutured to the right and left ventricles. The patches that provided the electrical contact for defibrillation required a thoracotomy for insertion. The most recent IDs are smaller, and can be placed in the chest in a subcutaneous pocket similar to a pacemaker unit. In addition, the lead system can be inserted transvenously, obviating the need for a thoracotomy and the morbidity and mortality associated. The device can be set to defibrillate when it senses a heart rate greater than a given predetermined limit. In addition, the device can be adjusted to trigger by assessment of the probability density function that evaluates time away from electrical baseline on the sensed ECG.

Early studies with the ID showed a remarkable response of very high risk patients. Lots of early shocks with a high survival supported the concept of a marked improvement in outcome (1). More recent studies have questioned the reasons for the shocks and the concept of appropriate (for life-threatening arrhythmia) and inappropriate shocks has developed (2). In addition, the substantial early mortality of implantation thoracotomy was not counted in survival, biasing the results in favor of the ID. Still, the body of evidence appears convincing that the ID, in very high risk patients, reduces the incidence of sudden death, although the effect on total mortality is not as dramatic (2). With the development of nonthoracotomy implantation that has essentially no associated mortality, combined with the new generation of devices with the ability to record the arrhythmia triggering the shock, the IDs appear more effective than ever. The days of extensive and prolonged serial drug testing are essentially over, when a device that is more reliable and better tolerated can be employed. Thus, antiarrhythmic drug therapy is often an additive therapy to a device. Still, there are extensive numbers of patients needing antiarrhythmic therapy for VT, VF, and supraventricular tachycardia (SVT) therapy. Which patients and what arrhythmias to treat are cogent issues that need to be addressed. Perhaps the best way to approach the problem is to discuss specific clinical arrhythmias and possible therapeutic approaches.

SUDDEN DEATH/CARDIAC ARREST SURVIVORS

The number of patients who survive a cardiac arrest is indeed small. Of the 300,000 to 400,000 sudden deaths each year, it is estimated that only about 2% survive the arrhythmic episode. The response time is slow in most urban as well as rural areas. The pharmacologic armamentarium employed may be inadequate and the ability to apply electrical shock therapy is very poor. This is a major problem area and one that is being addressed in a number of ways. There is a major effort to have low-cost automatic external defibrillators available for cardiac resuscitation in the field. In a collapsed, unconscious person, defibrillation is the first approach. Early defibrillation, CPR initiation, and rapid transfer to an ER are necessary to improve the survival statistics. Pharmacologic therapy has a role. The emphasis on lidocaine as first-line therapy has for too long permitted "placebo therapy" to remain a cornerstone. In the treatment of VT, lidocaine has an average efficacy of perhaps 8%, comparable to the success rate of placebo (3). Studies are now evaluating amiodarone in the IV Tween 80 and benzyl alcohol formulation Cordarone as well as the alternative investigational preparation Amio-Aqueous that can be given by bolus without the hypotensive effects of the Tween 80 and

benzyl alcohol (4). Uncontrolled studies suggest an efficacy rate of between 30% and 50% for IV amiodarone when used first line (5,6,7). IV amiodarone should be considered a first-line antiarrhythmic. If the patient has a pulse and arterial pressure, maintaining continuous, pharmacologic therapy can be tried first, followed in rapid succession by electroshock therapy. Often one, two, or three boluses of amiodarone are required combined with electroshock to terminate the VT/VF. It is not surprising if asystole ensues in patients who have been down for a period of time. Rapid temporary pacemaker insertion followed by pacing may be necessary to prevent asystole and death. Patients resistant to defibrillation may benefit from epinephrine administration, though success with this therapy is limited. IV lidocaine, bretylium, and procainamide, while available and having their advocates, are not as effective as amiodarone. The definitive answer for first-line therapy awaits the results of controlled randomized trials that are ongoing.

Once resuscitated, patients post arrest need to be thoroughly evaluated for etiology. Was the episode ischemic, due to hypokalemia, secondary to a drug toxicity, due to digitalis excess, or due to some other consideration? After remediable causes are evaluated and corrected, despite the presumed findings of an etiology, an electrophysiology (EP) study needs to be performed. It is sheer sophistry to find coronary artery disease (CAD) on cath in a middle-aged man and then attribute the cardiac arrest to CAD. The prevalence of CAD is high in the middle-aged male cohort and there need not be a direct cause-and-effect relationship. An EP study will evaluate the possible etiologies of VT, VF, SVT, or bradycardia. To do less than a complete EP study is less than the standard of care for cardiac arrest patients. While the Electrophysiologic Study Versus Electroradiographic Monitoring (ESVAM) suggested that Holter-guided therapy is as good as EP studies in defining therapy and an easier technique to find an effective agent, the results are limited and controversial (8). Both groups (EP and Holter) needed to be inducible at Programmed Electrical Stimulation (PES) and both groups needed frequent ambient arrhythmias. Most arrest patients don't have frequent VPCs and nonsustained VT and PES studies usually entail evaluating the patient with three extra stimuli. Certainly, a patient with an arrest and frequent ventricular arrhythmias can have antiarrhythmic therapy guided noninvasively. But the EP study is needed to evaluate the other possible concomitant problems such as SVT or conduction disturbance. Electrophysiologic evaluation on antiarrhythmic therapy is especially important.

It needs to be understood that the presence and extent of disease are critical considerations. Patients without coronary disease and a normal EF are distinctly different from the patients with structural heart disease, different as to etiology, drug efficacy, and the optimum selection of pharmacologic therapy. Patients with a low EF are at considerable risk for a recurrence, and survival is inversely proportional to EF. Patients with a low EF are more prone to proarrhythmia and are placed at greater risk by antiarrhythmic agents of the Class 1 and 3 variety (excluding amiodarone).

Thus, all patients with an arrest should undergo an electrophysiology study to evaluate automaticity, conduction, the possibility of SVT, preexcitation, and inducible VT or VF or noninducible VT. Those patients that are inducible are at higher risk. However, patients with an arrest but no clear etiology of the arrhythmia are still at risk, warranting aggressive therapy (9). Once risk is established, then an appropriate clinical treatment plan needs to be formulated. Given the good results with the third-generation implantable device, the ID must be considered first-line therapy. Still there are patients who prefer a pharmacologic alternative, or patients who have other severe disease and pharmacologic therapy is preferable to a device. Regardless of long-term therapy, a brief course of pharmacologic testing is indicated. The results will define potentially useful treatment(s) that can be used as stand alone or in combination with a device implantation. Procainamide is often advocated as a stand in for other agents, sort of a surrogate probe to see if other agents will work. However, there is no good evidence supporting procainamide as the surrogate. In fact, flecainide and procainamide, for example, show a lack of correlation of effectiveness as determined at EP testing (10). A logical testing program would include a Class 1c agent, sotalol, and amiodarone. EP testing on amiodarone is problematic. Horowitz and associates (11) reported a study suggesting that noninducibility or VT slowing is advantageous for patients on amiodarone. The real problem is the time required for oral loading. IV testing with flecainide, sotalol, and amiodarone is possible and time saving, though the lack of metabolites is a downside of this approach. Still, acute EP testing permits a good understanding of the conduction effects and the possibility of acute proarrhythmia, concerns best evaluated in the EP laboratory. In patients with EF lower than 35% or 40%, the use of a class Ic agent is best deferred.

It is estimated that between 30% and 50% of patients with a defibrillator implant are on an antiarrhythmic. Beta-blockers are used to prevent SVT or serious tachycardia that triggers the rate-sensing function for VT determination. In addition, the best IDs are those that are never needed. It is like a parachute worn on a test flight. The danger is there, the parachute is reassuring and needed, but it is best not to need to use the device. Appropriately guided antiarrhythmic therapy can offer protection with the further backup of the ID. What must be avoided are side effects from the antiarrhythmic. Antiarrhythmics can also cause a rise in defibrillation threshold, and this can interfere with the ID's effectiveness. Threshold testing is requisite when antiarrhythmic therapy is changed. While the percent change in defibrillator threshold may appear of a considerable magnitude, one infrequently encounters a clinically significant rise in threshold. With the new bidirectional waveforms and the "hot can" devices, the problem of a rise in threshold has little significance. Still, to avoid problems, evaluation for proarrhythmia and a change in threshold are needed, as well as assessment of LV function effects of the antiarrhythmic therapy. The negative inotropic effects of the available class Ic agents flecainide and propafenone are such that their use in low EF patients is to be avoided. Sotalol needs to be observed for changes in the QT. Prolongation beyond 440 ms is a warning sign for possible torsades de pointes. An increase in QT dispersion may also be an electrocardiographic warning sign of impending proarrhythmia. Fortunately, most proarrhythmia of this variety is reported to occur early in the course of treatment. However, with d-sotalol, the proarrhythmia was cumulative over the course of therapy (12). The problem with amiodarone is the protein noncardiac side-effect profile. Low-dose amiodarone therapy is indicated to avoid side effects, although the precise improvement in the side-effect profile with low-dose therapy has not been studied in controlled trials. A reduced loading regimen 800 mg a day for 4 to 6 weeks followed by 400 mg a day for a month and then 200 mg per day may be appropriate. Amiodarone levels of 1 µg/ml may be a very rough guide.

The new class III antiarrhythmic agents may be very promising for the high-risk patients with an ID needing rate control and a decrease in VT/VF incidence. The new agents are specific Ik_r blockers, with few extracardiac side effects. The torsades de pointes ventricular tachycardia this type of drug provokes seems to occur in the early loading stage and can be evaluated and excluded early in the course of therapy. As these agents become available, their place in therapy will become clearer.

The arrhythmia type being treated in the sudden death survivor affects the therapy selection. Monomorphic VT at PES Studies is very specific on PES studies and makes for a useful end point that is often reproducible. Drugs that suppress monomorphic VT have a high likelihood of clinical efficacy (75% to 85%). Polymorphic VT is a less specific marker. However, patients with a low EF are more likely to have polymorphic VT spontaneously and PES provoking polymorphic VT. Suppression of polymorphic VT by pharmacologic therapy is still a help to guide therapy selection. The use of an implantable device obviates the need for absolute specificity since what is being asked of the drug is a reduction in shocks and not 100% reliability.

HIGH-RISK PATIENTS FOR A FIRST EVENT

A much larger pool of patients are those at risk who have not had a cardiac arrest. How are these patients to be screened and how are they to be treated is a most controversial area and one of great importance. Appropriately treating this population will lead to meaningful results in sudden death prevention. But costs and the hope to avoid needless invasive testing and therapies limits what we can currently do. Patients post–myocardial infarction (MI) are at higher risk for sudden death. As EF decreases, VPCs increase, nonsustained VT develops, heart rate variability decreases, signal averaged ECGs become positive, QT dispersion increases, the exercise test is positive for ischemia early on, T wave alternans is present, and risk for sudden death increases. The Multicenter Automatic Defibrillator Implantation Trial (MADIT), a recent study of patients post-MI who have low EF with nonsustained VT and have sustained VT induced in the EP laboratory by PES techniques, but can't have the induction of the VT suppressed by procainamide, has found that an implantable defibrillator is significantly more effective than conventional therapy (75% amiodarone) in preventing sudden death (13). While MADIT II will evaluate if patients with a low EF (<40%) post-MI do better with a defibrillator, this result is suggested from the outcome of MADIT I.

In all probability all high-risk groups do better with a device than without. Patients with a dilated myopathy and low EF, patients postarrest, and patients with nonsustained VT and a substrate abnormality have lower sudden death mortality with an ID. What will be needed is a stripped-down, low-cost defibrillator making this strategy more economically practical. Currently, aggressive antiischemic therapy with a beta-blocker, combined with EP-guided class Ic or III therapy backed up in appropriate patients by a device, is appropriate therapy. The 2% to 3% of the post-MI population that fits the MADIT criteria should currently receive an ID as the standard of care.

VENTRICULAR FIBRILLATION THERAPY

The cornerstone of therapy for ventricular fibrillation is electrical defibrillation. In the acute setting, defibrillation is the first line of therapy. IV bretylium or epinephrine can occasionally affect a clinical conversion but this is infrequent and not to be depended on. Antiarrhythmia therapy is then adjunctive to defibrillation therapy. Patients resuscitated from ventricular fibrillation are at high risk of recurrence. Often at electrophysiologic testing monomorphic VT may be induced and the VT may be suppressible by antiarrhythmic therapy. The results at long-term follow-up are less successful in VF patients than VT patients. Thus, chronic therapy should employ the implantable defibulator (ID), preferably with concomitant supplementary antiarrhythmic therapy. Patients with VF and a normal heart are an uncommon but troublesome group. They are frequently noninducible at PES testing. Implantation of an ID is recommended since a good percentage of these patients are at risk for recurrence. Better strategies are needed to determine who is and who is not at risk for a recurrent life-threatening arrhythmia.

VT CHARACTERIZATION AND PHARMACOLOGIC THERAPY

There are a number of special situations that characterize the ventricular tachycardia and should be taken into account when deciding therapy. At times antiarrhythmic therapy may be directly determined by the arrhythmia being targeted for therapy.

ACCELERATED VENTRICULAR RHYTHM

Accelerated idioventricular rhythm (AIVR) or slow VT is often an escape rhythm that indicates underlying disease. Ischemia of the conducting system with the development of an accelerated rhythm of a subsidiary pacemaker may be responsible for the AIVR. Hence, therapy needs to be directed at the ischemia without the use of drugs that suppress automaticity and cardiac conduction. Early ICU experience has taught us not to employ lidocaine for AIVR since the result may be asystole. While AIVR may further accelerate and lead to hemodynamic decomposition, this is uncommon. The use of antiarrhythmic agents without the ability to insert a temporary pacemaker may place the patient at greater risk.

EXERCISE-INDUCED VT

Ventricular Tachycardia VT provoked by exercise has special therapeutic implications. The first priority is to determine if ischemia is the etiology of the VT. If indeed VT is a result of ischemia, a reversible thallium defect on nuclear stress test should be observed, though reproducibility of the VT is often poor. An interventional procedure (angioplasty or coronary bypass surgery) may be indicated. It is critical to repeat the exercise test after intervention to ensure VT prevention. Exercise VT can also develop due to autonomic imbalance (excessive sympathetic or marked parasympathetic withdrawal). The autonomic imbalance can cause dispersion leading to a reentrant tachycardia. There can also be abnormal automaticity triggered by sympathetic nervous system overstimulation, leading to a reentrant ventricular tachycardia. Often, the rhythm disturbance can be effectively treated by a beta-blocker or verapamil. A long-acting, once-a-day beta-blocker should be first-line therapy. After a steady state is reached through oral loading, the exercise test should be repeated to determine if the therapy prevents VT. Those patients who can't tolerate a beta-blocker should be started first on verapamil with an exercise test repeated. The new verapamil preparation (Covera) that is administered at night so that an effective concentration is obtained in the morning hours upon awakening theoretically appears attractive, since it would modulate heart rate acceleration as well as sympathetic facilitation of calcium mobilization that could precipitate calcium-triggered arrhythmias. Although less experience is reported with diltiazem, the drug appears effective based on antidotal clinical experience as well as less pressing negative inotrophy.

LQTS/TORSADE DE POINTES

The congenital long QT syndrome is uncommon but a unique situation in which molecular biology has explosively elucidated a syndrome that for over three decades was mired in misconception and therapies that were based on a misunderstanding of the underlying pathophysiology. Autonomic imbalance is not the etiology of the condition. Autonomic imbalance, though, can trigger arrhythmias in a subset of patients (LQT_1) and thus antiadrenergic therapy empirically administered can often be effective. The LQTS patients can be divided into LQT_1, LQT_2, and LQT_3 subgroups having chromosomal abnormalities on chromosome 3, 7, and 11, respectively. The abnormalities are heterogeneous and appear to be subtle and different from kindred to kindred. Patients with LQT_1 appear to benefit most from beta-blockers (long-acting recommended). LQT_2 patients may benefit from increasing K^+ serum concentrations (14). However, no long-term follow-up with K^+ therapy has been performed. LQT_3 patients may respond to sodium channel inhibition by mexiletine or other class Ib or Ia agents (15). The ability to genetically characterize the disorder, understand the ion channel abnormality and then select a specific therapy is a most exacting possibility

and one that many hope will be usable in other cardiac conditions as well. One recent report has related ECG findings to chromosomal abnormality and thus possible therapy selection (16). Presently, therapy of the congenital long QT should start with a beta-blocker, avoiding hypokalemia, and using mexiletine as adjunctive therapy in selective cases. Failure with this approach may then lead to ID implant. Alternatively, antiarrhythmics are usually ineffective and may be adverse (Class 1a in particular).

NONCONGENITAL LQT SYNDROME

Torsades de pointes ventricular tachycardia associated with a prolonged QT interval but not on an hereditary basis constitutes the noncongenital LQT syndrome. This syndrome may develop due to hypokalemia, hypomagnesium, a type Ia antiarrhythmic, especially quinidine, certain of the type III agents (sotalol or the pure Ik_r blockers) or other pharmacologic therapies. These other pharmacologic therapies are now known to block Ik_r the K^+ outward rectifier current that is involved in repolarization. Seldane (terfenadine) in slow metabolizers rises to concentrations that cause torsades de pointes (17). Other agents may change terfenadine metabolism that leads to torsades. Such agents are a ketoconazole or cyclosporine. Agents that directly act on the inward rectifier current and possibly directly on Ik_r are probucol, cisapride, erythromycin, and nortriptyline, to list some of the frequently mentioned agents. Therapy for the noncongenital syndrome should be aimed at withdrawal of the precipitating agent, with maintenance of a normal Mg^{++} and normal K^+ concentrations. Seldane (terfenadine) can be replaced by an alternative antihistamine like loratadine or the isomer of Seldane, Alegra, neither agent affecting Ik_r.

Acute therapy of torsades depends on the hemodynamic condition of the patient. Short bursts of torsades may be treated with IV magnesium and IV lidocaine followed by temporary pacemaker insertion to increase heart rate, decrease dispersion, and thus decrease or abolish the bursts of torsades. However, in patients with rapid hemodynamically decompensating torsades, electroshock followed by overdrive pacing is indicated.

BIDIRECTIONAL VENTRICULAR TACHYCARDIA

This is an unusual rhythm that is indicative of digitalis toxicity. The rhythm may be a harbinger of VF. The rhythm may be a premorbid one, and rapid treatment to prevent death by administration of the digoxin specific antibody is indicated. Digitalis toxicity infrequently manifests as double tachycardia. Digitalis toxicity presenting as nonsustained VT or runs of sustained VT can be as serious and is more frequently encountered than bidirectional VT. There are many ways to treat VT secondary to digitalis toxicity. Dilantin, betablockers, or lidocaine are effective therapies. One needs to be cautious with K^+ administration because as the digitalis toxicity becomes more severe, NaK adenosine triphosphatase (ATPase) inhibition increases and this may cause a rise in extracellular K^+. Exogenous K^+ combined with the rises in serum K^+ can combine to cause a decrease in automaticity and conduction resulting in asystole. Regardless of rhythm (double tachycardia, monomorphic VT, or nonsustained hemodynamically unstable VT) use of the digoxin-specific Fab frequently is the therapy of choice. Multiple antibody administration in close proximity for repeated toxicities is to be avoided to prevent immunologic reactions, and thus charcoal hemoperfusion is a viable alternative but one needing more time to be effective.

ARRHYTHMOGENIC DYSPLASIA

The condition of arrhythmic right ventricular dysplasia with left bundle branch block and sustained VT is a unique entity. From a review of the literature the condition appears more prevalent in Europe than in the United States. However, VT of this morphology in a young patient should raise an index of suspicion. The definitive test is a cardiac magnetic resonance imaging (MRI), which clearly shows the fatty infiltration of the right ventricle that is often thin and dystrophic. The ventricular tachycardia may be exercise related or, rarely, part of a congestive myopathy. A number of therapeutic options are available. Verapamil may be effective, especially in the exercise-related ventricular tachycardia (18). Sotalol therapy has been reported effective (19). For those patients not responding to medical therapy or those patients resuscitated from a cardiac arrest, implantation of an ID is most appropriate. Cardiac surgery involving disarticulating the right ventricle has been developed. ID placement requires less specific skill and expense than the surgery with a lower morbidity and mortality and thus has become a more frequent therapeutic choice. Anticoagulation to prevent pulmonary embolism in these patients may be necessary.

BUNDLE BRANCH REENTRY

In bundle branch reentry ventricular tachycardia the VT involves the right and left bundle branches as part of the tachycardia pathway. The arrhythmia is often associated with concomitant myocardial dysfunction (CHF, cardiac enlargement). The etiology may be ischemic or idiopathic dilated cardiomyopathy. The bundle branch reentry can be induced and identified in the electrophysiology laboratory. While antiarrhythmic therapy may be successful especially when guided by electrophysiology studies, endocardial catheter ablation of the VT origin is favored because ablation is definitive, not requiring indefinite medical therapy that can be associated with a proarrhythmic risk. Thus, if a bundle branch reentry VT can be identified, the standard of care is to perform ablation of the reentry pathway involving one of the bundles (20).

MONOMORPHIC VT WITH BUNDLE BRANCH BLOCK CONDUCTION

A specific tachycardia can be identified in patients without demonstrable structural heart disease that is monomorphic and is associated with a bundle branch block morphology. The tachycardia is not frequently associated with a cardiac arrest, although patients may present with syncope. Left ventricular (LV) function is normal in these patients. Right bundle branch block (RBBB) VT with left axis and normal LV function is often responsive to verapamil therapy, and this agent should be evaluated as a first-line choice for chronic therapy (21). Ablation therapy has been developed and in the skilled operator's hands is both highly effective as well as safe. A case can be made to use ablation first, sparing the patient a lifetime of medical therapy. The selection of therapy is often now based on the physician's skills in ablation or preference for medical therapy.

Left bundle branch block (LBBB) VT morphology with a right or normal axis locates the tachycardia reentry pathway; it usually involves the right ventricular outflow tract. Cardiac arrest is infrequent but syncope can occur. If the tachycardia is exertion related, verapamil or a beta-blocker may be effective therapy. Radiofrequency endocardial catheter ablation is an effective therapy for these types of patients. The arrhythmia is often characterized as an SVT with aberrancy and only a high index of suspicion combined with an electrophysiology study can permit the correct diagnosis to be made.

VENTRICULAR FLUTTER

Ventricular flutter should be looked upon as a very fast VT. Often the patient is hemodynamically decompensating and rapid electroshock is requisite to prevent death. Pacing termination may be tried but the rate of the V flutter is often too fast for pacing termination. The rhythm is at times associated with ischemia and LV enlargement and/or dysfunc-

tion. Optimization of therapy by EP testing does not appear different than sustained VT. However, the rapid nature of the flutter leads to repeated defibrillations in serial drug testing, and this alone may be reason to turn first to an implantable device.

SVT THERAPY

The treatment of supraventricular arrhythmia has undergone considerable change. The advances in SVT therapy have made the therapeutic armamentarium more complex with a far wider variety of therapies. With the difficulty of antiarrhythmic therapy showing a mortality benefit with ventricular arrhythmias, drug developers have turned to the SVT indication. The future will see even further therapy for SVT, making the choices of an informed clinician among an array of possibilities even more difficult than it is currently. A reasonable, evidence-based therapeutic decision-making process will become even more essential in the future.

There are pharmacologic agents for rate control, agents for acute conversion, and drugs for prophylaxis to prevent recurrence. These therapies are competing with electrophysiologic techniques to interrupt the reentry arrhythmia pathway. The development of the technique of radiofrequency ablation has rapidly changed SVT therapy. An example is seen with patients who have a preexcitation syndrome. We have gone from pharmacologic to surgical therapy and now the treatment of choice is radiofrequency ablation therapy. The evolution of the field has been quite rapid. Similar changes are arising with other conditions. For instance, with atrial fibrillation pharmacologic therapy is changing. The debate between rate control or conversion followed by antiarrhythmic drug prophylaxis is ongoing, with the first randomized trial to determine the preferable approach under way (22). Competing with the pharmacologic approach are surgical techniques such as the MAZE procedure or a catheter approach creating a similar end result as seen with the MAZE operation (23). Just as in the field of ventricular arrhythmias, devices are beginning to undergo investigation for the management of SVT. Antitachycardia pacing has been under study for a number of years with a variable success rate. An implantable device is just beginning trials by the company In Control that utilizes an algorithm to detect AF and then administer a very low energy, synchronized shock to terminate the arrhythmia. Could this device do to AF therapy what the implantable defibrillator has done for VT/VF therapy? The developments in therapeutics in areas dormant for many years are now both varied and fast. Exciting clinically useful advances are to be expected in the SVT field.

One very important concept regarding pharmacologic therapy is the association of low EF with increased proarrhythmia. A class Ic agent like flecainide or propafenone can be limited in its effectiveness in patients with structural heart disease and low EF. In compromised patients, deterioration in LV function and the facilitation of reentrant ventricular arrhythmias limit the use of class Ic agents in the treatment of ventricular arrhythmias. These same agents are very effective in SVT therapy but may be adverse in patients with structural heart disease with poor LV function. In addition, the interaction of ischemia and class Ic action combine to facilitate the development of wide complex ventricular tachycardia that can lead to incessant arrhythmias and death. This may be the underlying mechanism of the increased mortality seen in the CAST trial. Thus, patients with SVT can be treated effectively with the class Ic agents or they can be placed at risk by them. It is the presence of LV dysfunction, structural heart disease, or ischemia that increases the potential for adversity from therapy with the class Ic agent. With this concept in mind class Ic therapies can be used effectively in patients with normal hearts and minimal disease. Utilizing exercise testing to screen for ischemia and the development of wide complex tachycardia may further increase the safety margin when employing class Ic agents (24).

As with ventricular tachycardia therapy an arrhythmia-specific, evidence-based therapeutic decision-making approach can be useful for the clinician. Therapeutic options by rhythm disorder is discussed in the following sections.

ATRIAL PREMATURE CONTRACTION THERAPY

The therapy of atrial premature beats depends on the patient's complaints. Asymptomatic APCs need not be treated. However, their presence should suggest further evaluation. Is their occult valvular disease, coronary artery disease (CAD), recent heart failure or pericardial disease? APCs in the context of an enlarged atrium may be a harbinger of AF. Asymptomatic APCs may represent an increased risk for AF and suppression of APCs may be useful in prolonging the time to AF. Class II agents may be effective in APC suppression and in slowing the ventricular response when AF occurs. The use of class Ic agents has not been studied adequately in this context to assess the risk-benefit ratio of these compounds. Previously, digitalis has been utilized in this situation, but the vagal effects of digitalis on atrial refractoriness favor the development of AF and should be avoided. In patients with recent congestive heart failure (CHF) and the potential for AF, digoxin therapy facilitates the development of atrial fibrillation. In light of the inefficiency of digitalis therapy in the Dig Trial (25) its profibrillatory action suggests that it should be avoided except in symptomatic heart failure patients. Patients with symptomatic APCs may be effectively treated with beta-blockers. Those intolerant to beta-blockers (a considerable number) may be treated by verapamil or diltiazem. The once-a-day preparations offer heightened patient compliance but may not maintain electrophysiologic action adequately around the clock, making use of the immediate release preparations necessary. A circadian release system may be potentially useful to blunt the early AM acceleration seen with sympathetic tone.

SINUS TACHYCARDIA

There are indeed times that sinus tachycardia needs to be treated. Sympathetic tone may be excessive and this can cause lasting myocardial dysfunction. Patients with inappropriate persistent sinus tachycardia may benefit from beta-blocking therapy. Persistent sinus tachycardia left untreated can cause a cardiomyopathy and in fact is used to create cardiomyopathies in animals (26). Sinus tachycardia may also be adverse in a similar way in patients with CHF. The clinician addresses the underlying cause of the tachycardia, but often, regardless of therapy, the heart rate can't be reduced in CHF patients. Studies have found that CHF patients with persistent sinus tachycardia benefit the most from beta-blockade (27). Slow introduction of the beta-blocker is key to successful therapy. Initial dose of 1 to 5 mg of metoprolol using pediatric suspension may be appropriate. The newer vasodilating beta-blockers that possess alpha-blocking action such as carvedilol and betaxolol may offer advantages since the afterload reduction component permits better tolerance of the beta-blocking effects that decrease LV function. However, could one not attain the same effect with aggressive angiotensin-converting enzyme (ACE) inhibition combined with beta-blockade? The advantage of two agents is that each can be titrated independently. With carvedilol and betaxolol, the beta-blocking dose is limited by the vasodilatation and ensuing fall in arterial pressure. As the patient with heart failure tolerates the beta-blocker over 2 to 3 months, further benefit from greater beta-blockade could be obtained but is often limited due to the depressor effect of further titration of the vasodilating beta-blockers. Using two drugs may afford the physician a greater range of sympathetic tone reduction with less vasodilation. The downside of two agents is compliance and possible increased cost.

ATRIAL ECTOPIC TACHYCARDIA

Atrial ectopic tachycardia may be nonsustained or sustained. The arrhythmias at times cause symptoms usually due to increased rate. Sympathetic autonomic tone is a facilitator of the arrhythmia. Thus, effective treatment is often the use of beta-blockers or, in the patient intolerant to beta-blocker therapy, verapamil or diltiazem can be utilized.

PROXIMAL ATRIAL TACHYCARDIA

Proximal atrial tachycardia (PAT) is a common atrial arrhythmia having an atrial rate of 150 to 250 that is either terminated or unaffected by increased vagal stimulation. The pathoelectrophysiology can consist of an intraatrial reentry tachycardia, an AV nodal tachycardia, and AV reciprocating tachycardia as well as an ectopic atrial tachycardia. An acute presentation of what appears to be an SVT often needs diagnostic confirmation followed by therapy. IV adenosine has become a mainstay of diagnostic therapy often leading to arrhythmia termination as well. While patients are symptomatic the arrhythmia is usually not fast enough to cause hemodynamic instability and thus the arrhythmia does not need acute electroshock intervention. Diltiazem or verapamil can be employed IV for termination. Verapamil IV is less costly but possessing more LV suppressing effect. It is important to exclude a preexcitation etiology if digoxin is to be used. In addition, in patients with a rapid tachycardia and in whom an AV reciprocating tachycardia as part of a preexcitation syndrome is possible, the patient may experience a serious episode of hypotension caused by verapamil or diltiazem, which precludes the use of these agents. IV procainamide is the agent of choice administered 100 mg every 5 minutes to a total of 1 g followed by an IV infusion of between 2 and 6 mg/min.

Chronic prophylaxis can include the use of digoxin, a beta-blocker, diltiazem, or verapamil. In persistent cases in patients with a normal LV, flecainide is especially effective. Patients with preexcitation are usually young, and a lifetime of antiarrhythmic therapy is problematic. Also, digitalis glycosides are to be avoided in preexcitation patients. The class Ic agents when found effective at EP testing are well tolerated and effective as chronic prophylactic therapy. However, the high success rate and very low mobility and essentially no mortality has made radiofrequency ablation the treatment of choice for patients with the preexcitation syndromes. All patients with AF and preexcitation should be studied at electrophysiology testing and ablative therapy considered. Surgical procedures, while effective, are no longer needed in the overwhelming majority of cases given the high success rate and very low adversity seen with ablative therapy.

Recurrent PATs as part of the "holiday heart" syndrome are most effectively treated with beta-blockers since often the alcohol withdrawal state leads to hyperadrenergic tone that precipitates arrhythmias. In patients with low EF who don't tolerate beta-blocker therapy and who may have an implanted device for VT, recurrent PATs can trigger the ID. These patients often don't tolerate class Ia and Ic agents and they are best treated with amiodarone. Low-dose chronic oral amiodarone therapy is most effective with a reduced incidence of toxicity chronically (28).

MULTIFOCAL ATRIAL TACHYCARDIA

The arrhythmia multifocal atrial tachycardia is often associated with pulmonary disease and in the end stages of severe chronic obstructive pulmonary disease (COPD) can be a problematic arrhythmia difficult to treat effectively. Verapamil is helpful and should be the first-line therapy. For those patients not responding to verapamil, amiodarone therapy is indicated. There is no evidence that the adverse side effect of pulmonary fibrosis due to amiodarone is increased in patients with underlying pulmonary disease. Thus, pulmonary disease is not a reason to avoid amiodarone therapy.

ATRIAL FIBRILLATION

Atrial fibrillation (AF) is a common rhythm, with 2% of the population experiencing this rhythm disturbance. For patients over the age of 65 the prevalence of AF increases to 6%. As our population ages, AF is becoming a very frequently encountered arrhythmia. Until recently, few controlled clinical trials were performed in patients with AF. The adversity of embolization risk has been known for years, but in addition to this problem, AF activates the ventricle irregularly, which increases electrical dispersion and predisposes the ventricle to reentry, arrhythmias, and sudden arrhythmic death. Still, we do not know if it is better to remain in normal sinus rhythm (NSR) or to obtain rate control of the AF. Given the problems of recurrent emboli, facilitation of arrhythmic death and the development of cardiomyopathy due to poor rate control maintenance of sinus rhythm is to be preferred if it can be obtained with a reasonable therapeutic regimen.

The acute management of AF can entail rate control or conversion. Those in hemodynamic compromise, those with angina, and those patients not responding to rate control are best treated with electrical cardioversion. Patients should not be cardioverted if they have been in AF more than 48 hours and not anticoagulated unless it is an emergency (patient in shock or CHF). Chemical conversion has been reported recently to be less associated with atrial contractile paralysis than electroshock (29). However, the risk of proarrhythmia with pharmacologic conversion is higher. To prevent emboli in patients in sinus rhythm but with poorly or noncontracting atria, anticoagulating with heparin followed by 2 to 4 weeks of coumadin is indicated. As with all anticoagulation with coumadin an international normalized rate (INR) of 1.5–2 is adequate and reduces the risk of bleeding. Ibutilide is very effective in causing acute conversion to NSR, though the high incidence of torsades of 5% or 6% and ibutilide's cost may limit its use. Flecainide is effective in acute conversion but no IV drug formulation is available and loading with the oral dosage formulation in patients with LV dysfunction is not recommended due to proarrhythmia. Both amiodarone and sotalol are not very effective in termination of AF, though they are very useful in preventing recurrence in patients postcardioversion. Amiodarone is especially useful in patients with a low EF due to the low to nonexistent incidence of proarrhythmia. The IV preparation available, Cordarone, does have the problem of hypotension and negative inotropy due to the excipients Tween 80 and benzyl alcohol. Slow administration of IV Cordarone may avoid these problems, or the experimental aqueous preparation Amio Aqueous can be employed to advantage.

The priority in AF acute therapy is to terminate the arrhythmia if the patient is in distress. If the patient is stable, then rate control with diltiazem, verapamil, or IV amiodarone is indicated. Following stabilization, conversion can be done electrically. If the onset of the AF is unknown or greater than 48 hours, anticoagulation is indicated. Use of transesophageal echo (TEE) is often not very helpful and is difficult to obtain if the patient needs urgent cardioversion. Digoxin therapy is to be avoided due to its profibrillatory action through the autonomic nervous system and its effect to decrease atrial refractoriness.

Patients with intermittent proximal AF and normal LV function without ischemia are best treated with a class Ic agent. Flecainide appears the most effective and the best tolerated. Those patients with low EF, ischemia, or recent history of an MI are best treated with low-dose amiodarone, which possesses a high efficacy and low proarrhythmia.

The chronic maintenance of sinus rhythm can be obtained with pharmacologic therapy. The class Ic agents have the highest efficacy, 70% to 90%, but their use is limited by proarrhythmia. Sotalol and amiodarone have lower efficacy, 50% on average, but can be given to the low-EF patient. Sotalol has the problem of producing torsades but has associated less severe systemic toxicities than amiodarone. Low-dose amiodarone offers the possibility of less significant systemic toxicities. The implantable cardioverter and the MAZE or Guirardom procedures should be reserved for the most recalcitrant of cases. Atrioventricular (AV) node modification in patients who can't maintain NSR is a viable therapy. With the modification technique permanent pacing can be avoided in the majority of cases. Long-term anticoagulation following ablation or surgery is indicated as well as in all patients with paroxysmal AF, persistent AF maintained at a slow rate by pharmacologic therapy, and patients with known atrial thrombus. If a patient is reverted to sinus rhythm, anticoagulants must be maintained for 4 weeks due to the possibility of late embolization that occurs when the atria begin to contract. In patients with intermittent AF or persistent AF and no structural heart disease, lone AF may be treated with aspirin therapy. There must be no history of arteriosclerotic heart disease (ASHD), EF must be normal, and the patient should be less than 60 years of age without a history of hypertension.

Despite the therapeutic armamentarium and surgical procedure now available, AF often cannot be prevented. Perhaps our target should be increasing the interval between recurrences. In those patients with AF recurrence, a repeat trial of NSR if the left atrial dimension is less than 6 cm is indicated. Otherwise, rate control on a chronic basis will be required. Digoxin is inadequate for rate control since sympathetic tone increases AV node conduction, overriding the direct effect of digoxin on the AV node. Diltiazem, verapamil, beta-blockers, or amiodarone are better choices than digoxin for rate control.

Recent evidence supports the use of a dual-chamber sensing and pacing DDD pacemaker. Atrial pacing prevents the later development of AF, while ventricular pacing favors the development of AF. Give the sequelae of AF, the advantages of DDD pacing favors its use despite the initially higher cost.

ATRIAL FLUTTER

Atrial flutter is associated with atrial rates usually between 250 and 350 beats per minute. There is the common type 1 flutter with inverted flutter waves seen in the inferior leads (II, III aVF). Pacing termination is often possible with type 1 flutter. Type 2 flutter is less common, has a faster rate, often between 350 to 450 atrial depolarizations per minute, and cannot be terminated by pacing.

Therapy first aims at rate control. Those in hemodynamic distress are best terminated by cardioversion. Stable patients can have rate control established with IV diltiazem, verapamil, or IV amiodarone. Often, the acute flutter will transition through atrial fibrillation before being terminated pharmacologically. Chronic prophylaxis of flutter employs the same agents as discussed with AF. Class Ic agents can be used but with concerns about proarrhythmia. Sotalol and amiodarone are effective, and therapy can be given for recurrent atrial flutter. Catheter ablation especially in the region around the isthmus between the inferior tricuspid ring and the orifice of the inferior vena cava can be employed. Operative therapy is possible but the ablation technique is preferable. Atrial flutter may not need chronic anticoagulation in light of atrial contraction and lack of blood stasis. Echocardiography may aid in the determination of atrial contraction, obviating the need for chronic anticoagulation.

AV NODE REENTRY

The curious rhythm of a tachycardia with two populations of PR intervals is now known as AV node reentry with dual pathway in the AV node permitting reentry. There is a slow and a fast pathway that is anatomically different at the junction. The onset is sudden, usually with no precipitating factors, and the symptoms most frequently reported as palpitations, though syncope may occur. Acute therapy may consist of adenosine, vagotonic maneuvers, or cardioversion. Chronic drug therapy may be directed at the anterograde limb of the circuit (class II or IV action) or the retrograde limb of the circuit (class Ic agents). Ablation of the slow pathway has become the therapy of choice with reports of a 90% success rate and less than a 1% incidence of AV block (30,31).

JUNCTIONAL TACHYCARDIA

Nonparoxysmal AV node tachycardia may be seen in the pediatric population as well as the adult patient. Both rhythms are responsive to beta-blocker therapy or IV amiodarone. The arrhythmia in adults can be seen in the acute MI period, at cardiac surgery, and due to digitalis intoxication. If digitalis toxicity is a possibility, stopping digoxin may be sufficient therapy.

SUMMARY

Many approaches to therapy are possible. There are far more agents available and approaches than a clinician needs to use. However, a few agents stand out because of efficacy, ease of use, and safety. In general, class II and IV agents are preferred for the reasons of low toxicity. While class II action is theoretically very useful, patients often can't tolerate a class II agent, and the class IV agents verapamil and diltiazem are employed. The class Ia agents are becoming of historical interest except for the use of IV procainamide due to its ease of administration and acute safety. Class Ic agents are effective but the problem of proarrhythmia and LV dysfunction limit their use. Flecainide appears to be the most effective agent of this group and the best tolerated. Within the class III group amiodarone is a uniquely effective agent. The potential for proarrhythmia is very low. While the drug prolongs the QT, it decreases QT dispersion and the development of torsades is rare (and when studied carefully, probably nonexistent) (32). In fact, amiodarone has been given to patients with torsades and long QT as therapy (33). The limitations with amiodarone are the protein side effects, the most serious being pulmonary fibrosis. Low-dose amiodarone appears to be less prone to produce these adversities, but the safety of low-dose amiodarone has not been confirmed in a controlled trial. While some turn to sotalol due to a shorter half-life and less systemic side effects, the efficacy of sotalol is less than amiodarone and the proarrhythmia with torsades is a problem. The newer agents with type III action appear similar to sotalol without beta-blocking activity. Torsades to a greater or lesser degree remains a problem. Why amiodarone retains such a unique place in therapy after 30 years is an enigma for the field of antiarrhythmic therapeutics. Amiodarone decreases dispersion when other Class 3 agents often increase dispersion, but the fundamental mechanisms behind these effects and profiles remain unexplained.

Nonpharmacologic therapy has undergone considerable advancement. The recent report of the premature stopping of the Amiodarone Versus Implantable Defibrillator (AVID) study with a 30% reduction in mortality in patients with an implantable defibrillator as compared to drug therapy further supports the use of defibrillators in the long-term management of patients with cardiac arrest. How to apply the AVID results outside of the cardiac arrest survivor group and to the large populations of patients with MI and CHF that are at risk is a substantial problem. Pharmacologic therapy has seen the number of drugs expanded and its use refined based on the results of controlled clinical trials. Still, we are a long way from a therapeutic approach that is proven to reduce mortality in a significant percentage of the patients at risk. This is the challenge for the next decade and it will be met.

REFERENCES

1. Echt D, Armstrong K, Schmidt P, Oyer P, Stinson E, Winkle R. Clinical experience complications and survival in 20 patients with the automatic implantable cardioverter/defibrillator. *Circulation* 1985;71:289.

2. Kim SG, Fisher JD, Furman S, et al. Benefits of implantable defibrillators are overestimated by sudden death rates and better represented by the total arrhythmic death rate. *J Am Coll Cardiol* 1991;17:1587.

3. Nasir N Jr, Taylor A, Doyle TK, Pacifico A. Evaluation of intravenous lidocaine for the termination of sustained monomorphic ventricular tachycardia in patients with coronary artery disease with or without healed myocardial infarction. *Am J Cardiol* 1994;74:1183.

4. Gallik D, Meissner M, Singer I, Hanin L. Hemodynamic and electrocardiographic effects of Amio-Aqueous: a new formulation of intravenous amiodarone [abstract]. NASPE, 1997 (in press).

5. Morady F, Scheinman M, Shen E, Spaira W, Sung R, Dicardo L. Intravenous amiodarone in the treatment of recurrent symptomatic ventricular tachycardia. *Am J Cardiol* 1983;51:156.

6. Somberg, JC. Intravenous Amiodarone. *PACE* 1986;4-5:436.

7. Scheinman MH, Levine JH, Cannom DS, et al. Dose-ranging study of intravenous amiodarone in patients with life-threatening ventricular tachyarrhythmias. *Circulation* 1995;92:3264.

8. Mason J for ESVEM. A comparison of seven antiarrhythmic drugs in patients with ventricular tachyarrhythmias. *N Engl J Med* 1993;329:452.

9. Crandall BG, Morris CD, Cutler JE, et al. The implantable cardioverter-defibrillator in noninducible sudden death survivors [abstract]. *Circulation* 1991; 84(4):609.

10. Wynn J, Torres V, Flowers D, et al. Antiarrhythmic drug efficacy at electrophysiology testing: predictive efficacy of procainamide and flecainide. *Am Heart J* 1986; 111:632.

11. Horowitz LN, Greenspan AM, Spielman SR. Usefulness of electrophysiologic testing in evaluation of amiodarone therapy for sustained ventricular tachyarrhythmias associated with coronary heart disease. *Am J Cardiol* 1985;55:367.

12. Waldo AL, Camm AJ, deRuyter H, et al. Effect of d-sotalol on mortality in patients with left ventricular dysfunction after recent and remote myocardial infarction. *Lancet* 1996;348:7.

13. Moss A, Hall J, Cannon D, et al. Improved survival with an implanted defibrillator in patients with coronary disease at high risk for ventricular arrhythmia. *N Engl J Med* 1996;335:1933.

14. Compton S, Lux R, Ramsey M, et al. Genetically defined therapy of inherited long QT syndrome: correction of abnormal repolarization by potassium. *Circulation* 1996;94:1018.

15. Schwartz PJ, Silvia G, Priori MD, et al. Long QT syndrome patients with mutations of the SCN5A and HERG genes have differential responses to Na^+ channel blockade and to increases in heart rate. *Circulation* 1995;92:3381.

16. Moss A, Zareba W, Benhorin J, et al. EKG T-wave patterns in genetically distinct forms of the hereditary long QT syndrome. *Circulation* 1995;92:2929.

17. Ray M, Dumaine R, Brown A. HERG, a primary human ventricular target of the nonsedating antihistamine terfenadine. *Circulation* 1996;94:817.

18. Woelfel A, Foster JR, McAllister RJ Jr, Simpson RJ, Gettes LS. Efficacy of verapamil in exercise induced VT. *Am J Cardiol* 1985;56:292.

19. Wichter T, Borggrefe M, Haverkamp W, Chen X, Breithardt G. Efficacy of antiarrhythmic drugs in patients with arrhythmogenic right ventricular disease. Results in patients with inducible and noninducible ventricular tachycardia. *Circulation* 1992;86:29.

20. Toubol P, Kirkorian G, Atallah G, et al. Bundle branch reentrant tachycardia treated by electrical ablation of the right bundle branch. *J Am Coll Cardiol* 1996;7:1404.

21. Thakur RK, Klein GJ, Sivaram CA, et al. Anatomic substrate for idiopathic left ventricular tachycardia. *Circulation* 1996;93:497.

22. AFFIRM Investigators. The design of the AFFIRM study. *Am J Cardiol* 1997;in press.

23. Haissagrevve M, Gencel L, Fischer B, et al. Successful catheter ablation of atrial fibrillation. *J Cardiovasc Electrophysiol* 1994;5:1045.

24. Ranger S, Talaji M, Lemery R, Ray D, Nattel S. Amplification of flecainide-induced ventricular conduction slowing by exercise. *Circulation* 1989;79:1000.

25. The Digitalis Investigation Group. The effect of digoxin on mortality and morbidity in patients with heart failure. *N Engl J* Med 1997;336:525.

26. Wijffels M, Kirchhof C, Dorland R, Allessi M. Atrial fibrillation begets atrial fibrillation: a study in awake chronically instrumented goats. *Circulation* 1995;92:1954.

27. Kubler W, Haass M. Cardioprotection: definition and fundamental principles. *Heart* 1996;75:330.

28. Kerin NZ, Aragon E, Faitel K, Frumin H, Rubenfire M. Long-term efficacy and toxicity of high- and low-dose amiodarone regimens. *J Clin Pharmacol* 1989;29:418.

29. Aranda JM, Tauth J, Brewington J, Lockeby M, Fontanet H. Atrial stunning occurs after electrical cardioversion, but not after chemical cardioversion [abstract]. *J Am Coll Cardiol* 1997;2:292A

30. Lee MA, Morady F, Kadish A, et al. Catheter modification of the atrioventricular junction with radiofrequency energy for control of atrioventricular nodal reentry tachycardia. *Circulation* 1991;83:827.

31. Jackman WM, Beckman KJ, McClelland. Treatment of supraventricular tachycardia due to atrioventricular nodal reentry by radiofrequency catheter ablation of slow-pathway conduction. *N Engl J Med* 1992;327:313.

32. Hii JTY, Wyse G, Gillis AM, Duff HJ, Solylo MA, Mitchell LB. Precordial QT interval dispersion as a marker of torsades de pointes. Disparate effects of class Ia antiarrhythmic drugs and amiodarone. *Circulation* 1992;86:1376.

33. Rankin A, Pringle S, Cobbe S. Acute treatment of torsades de pointes with amiodarone: proarrhythmic and antiarrhythmic association of QT prolongation. *Am Heart J* 1990;119(1);185.

12

Therapeutic Perspective

E. M. Vaughan Williams

Nature should be left to do her own work, while the physician stands by, as it were to clap her on the back, and encourage her when she does well.

Henry Fielding, *The History of Tom Jones*

SCIENTIFIC CARDIOLOGY

In spite of his skepticism Fielding optimistically believed that "if the numbers of those who recover by physic could be opposed to that of the martyrs to it, the former would rather exceed the latter." With hindsight his judgment would appear generous, since "physic" involved blood letting, administration of mercury and arsenic, and the transport by the doctor, like a bee laden with lethal pollen, of infection from bed to bed. Indeed, it was revulsion against current medical practice that induced Hahnemann, who was born a year after Fielding's death (1754) and who became a disciple of von Quarin, Emperor Joseph's physician, to found his system of homeopathy, which permitted Nature to do her own work without interference. The flourishing of physiologic science in the 19th century was but slowly transferred to medicine, and cardiology as a scientific discipline is less than a century old. The first journal devoted to cardiovascular diseases, *Maladies du Coeur et des Vaisseaux*, started publication in 1908, followed in 1910 by Lewis's *Heart*, a title to which the *British Heart Journal* has now reverted. The *American Heart Journal* was founded in 1926.

Even after the Second World War there was no cardiac surgery, which had to await the development of extracorporeal circulation, and there were few antiarrhythmic drugs. Rigorous testing of new remedies and procedures was accelerated by the disasters of thalidomide and practolol, and is still under review. Thalidomide was an effective and safe hypnotic in adults, but was found to be teratogenic and was not withdrawn until a large number of babies had been born with severe limb deformities. Practolol was the first of the selective β_1-adrenoceptor blockers, but when toxicity to serous membranes was observed in a small number of patients, it was withdrawn immediately. As the next millennium approaches we cannot guarantee to follow the precept *primum non nocere*, and we have been forced to admit that treatment may sometimes be harmful, as was demonstrated by the Cardiac Arrhythmia Suppression Trial (CAST). "When tragedy occurs (and CAST should be thought of as no less than a tragedy) we can only hope to learn" (1).

FAITH AS HEALER

The remarkable and continuing advances in cardiology are undeniable but expensive. Coronary artery bypass surgery or angioplasty, implantation of pacemakers, and lifelong administration of costly drugs are limited to the lucky or the loaded in the affluent societies of the world. Even among those to whom it is available, however, there is often a reluctance to accept modern medical treatment. Roman emperors pandered to the animal in their subjects by providing "bread and circuses," the latter including gladiatorial fights to the death or the spectacle of humans being devoured by wild beasts. The thirst for blood is now sated by violent films or scenes of operations as entertainment. The impotent patient lies prostrated upon the table, festooned with tubes and surrounded by flashing screens, while the surgeon hovers over him with a scalpel, like an Inca priest poised to perform a human sacrifice, ready to plunge the blade into his bleeding flesh and reveal his entrails in gory detail. Thus indoctrinated, many patients delay consultation, both from hesitation to incur expense and for fear of being cast in the role of victim.

In parallel with advances in medical technology there has been a growth of obscurantism and alternatives to orthodox medicine. A demand for miracle cures finds suppliers eager to relieve the gullible of their anxieties and their cash by claiming access to the healing power of meditation, ritual exercises, of various vegetable or animal extracts, perhaps diluted to an innocuous nth degree, or of direct intervention by the divine. There is no shortage of devotees willing to bear witness to personal benefit, and though such claims may be ridiculed, they cannot be ignored. Can faith in the healer have a therapeutic effect, and influence the outcome of a trial?

The introduction of double-blind, placebo-controlled investigations, sometimes with crossover of treated and untreated groups, in order to eliminate bias in patient or doctor, implies an acceptance of the possibility that knowledge of the treatment could influence the result. It has often been believed that a strong desire to be cured, and the will to live, can in some way fortify the body's response to disease. If true, the double-blind procedure might actually eliminate a benefit. Faith in the power of mind over body is common, but objective evidence is not easily acquired. In a recent study of prognosis and risk stratification in patients after a myocardial infarction (MI) it was concluded that "major depression in patients hospitalized for acute MI has been shown to be an independent risk factor for mortality at 6 months. Persistent postinfarction depression was an independent and important

source of subsequent morbidity and long-lasting reduced quality of life" (2). Exercise and physical activity are of recognized benefit in patients with ischemic heart disease, though it has been suggested that improvement in morale may be largely responsible for the effect. There is also objective evidence that exercise training can influence autonomic control of heart rate in patients with heart failure, directing the 24-hour variation in circadian rhythm toward a more normal pattern (3).

If faith in the treatment or its administrator is beneficial, the question arises how to encourage this without sacrifice of scientific objectivity. Legal responsibility for failure and the threat of litigation increases the cost of conventional medicine by imposing a burden of insurance not applicable to "alternative" advisers who peddle their wares. A single error by a qualified practitioner attracts more publicity than a thousand successes. To return to Fielding, he recounts a visit by two doctors to a patient, a naval captain, who had unfortunately passed away before their arrival.

"Every physician has his favourite disease. The learned gentlemen, instead of endeavouring to revive the patient, fell into a dispute on the occasion of his death. The Captain had been put into a warm bed, had his veins scarified, his forehead chafed, and all sorts of strong drops applied to his lips and nostrils. The physicians, therefore, finding themselves anticipated in everything they ordered, were at a loss how to apply that portion of time which is usual and decent to remain for their fee."

A cynic would complain that little has changed; arrogance and cupidity are not unknown today, and the incompetence of a few may drive many into the clutches of quacks and charlatans, whose malign propaganda can be countered only if orthodox treatment reliably improves the quality and duration of life, which cannot be guaranteed.

DEVELOPMENT OF NEW DRUGS

Molecular biology now dominates research. At the end of another century of scientific cardiology the human genome will probably have been unraveled, and the detailed structure of every protein will have been elucidated. The exact site of attachment of every drug will be known, and its role in modifying cellular function will be understood. At present, however, it is difficult to extrapolate from studies of drug interactions with receptor molecules to clinical effects, and experiments on whole animals or at least on isolated but still functioning organs are indispensible. Conservationists, who recognize that homo sapiens is rapidly polluting the planet until the environment may be able to sustain only such hardy survivors as the cockroach or horseshoe crab, may wish to protect all species equally and vigorously oppose the use of animals for research, but must recognize that if they succeed, medical advances cease.

Extrapolation of results even from mammals to man is insecure, and requires at least that a range of concentrations of a new drug be studied at levels that embrace a clinically usable range. For example, in a recent investigation (4), beta-blockers were administered to rats at 80 mg/kg, at least ten times the clinical dose, so that no valid conclusions could be drawn from effects observed. Ultimately, new drugs must be tested on human volunteers and consenting patients. A recent study in healthy young males who underwent cardiac catheterization (with one incident of atrial fibrillation, necessitating anesthesia and countershock) was much criticized as unethical by some, but praised by others. Recruitment for the armed forces is encouraged by all nations, and is compulsory in most. If it is acceptable to risk death in learning or attempting to kill other humans, who has the authority to deny to an idealist the opportunity to undertake a small risk in a project that might be of benefit to all?

When drugs are tested in patients, it must somehow be decided whether the trial is a success or failure. The clearest unsatisfactory outcome is death, but since the only certainties in this world are death and taxes, if the trial is sufficiently prolonged, 100% of outcomes will be unsatisfactory. Mortality can be recorded at intervals, and comparisons made between treated and untreated groups. Though statistically secure, the procedure ignores other pos-

sible surrogate end points, such as reinfarction, lowering of blood pressure or heart rate, incidence of stroke, etc. A drug was recently licensed on the basis that it was proven to lower blood cholesterol in man, which implied the acceptance of the unproven assumption that it would also improve the quality and/or the duration of life (5). Recent trials of the statins support the view that the assumption is valid. Many patients refuse further treatment because they suffer unacceptable side effects, deciding that, even if persuaded they might live longer, they would prefer to eat, drink, and be merry during their remaining days on earth. Conversely, should a drug be licensed that subjectively improves the quality of existence, even though there is evidence (as with some positive inotropic remedies) that mortality is higher than in patients on placebo?

Deciding what evidence to accept as a surrogate end point for success or failure presents an unsolved problem both for physicians and licensing authorities (5). A recent paper suggests that appropriately designed questionnaires filled in by patients may be more reliable as evidence of improved quality of life than objective tests of exercise tolerance on a treadmill (6). The authors concluded, "Our insistence on objective measures must be rethought, and the inclusion of quality-of-life or health status questionnaires in future studies must be encouraged. It would, therefore, seem sensible to develop and validate questionnaires which are disease specific. This whole area is clearly important in all patients with diseases that cause symptoms, and can no longer be ignored."

What profiteth a man to manage a minute more on a treadmill if he feels miserable meanwhile?

SELECTION OF PATIENTS FOR ANTIARRHYTHMIC THERAPY

Multiple physiologic feedback controls ensure that a regular and appropriate cardiac rhythm is maintained night and day, but they are not infallible, and normal healthy individuals experience a variety of cardiac arrhythmias. In a study of 440 normal subjects episodes of first-degree atrioventricular block were observed in 17.4%, and second-degree block in 10.5% (7). Isolated atrial premature beats (Pbs) occurred in 65%, and ventricular Pbs in 45%. Even couplets and multifocal arrhythmias were occasionally seen. How then to identify patients with arrhythmias of such severity that treatment is required? In the past the Lown classification was usually accepted as stratifying the degree of risk, but this has become less secure in the light of later studies, including that of Campbell et al. (8) on the natural history of acute MI. In a recent review Campbell (9) commented with reference to the Lown classification:

"The scheme arranged ventricular arrhythmias in a hierarchical and initially mutually exclusive manner. One suggestion, based on evidence derived from observations of arrhythmic events in the early stage of acute myocardial infarction, was that increasing "severity" of ventricular arrhythmias was associated with a risk of ventricular fibrillation (VF). We now know that preceding ventricular arrhythmia patterns do not predict VF. Although use of the Lown classification is diminishing, it has focused attention on the remarkable diversity of ventricular arrhythmias—isolated infrequent ventricular ectopic beats (VEBs), frequent uniform VEBs, multiform (varying QRS shape) VEBs, nonsustained ventricular tachycardia (VT), sustained VT, and VF."

What has emerged from many trials is that the most consistent predictor of cardiac death is the state of the myocardium, as revealed by the ejection fraction (EF). Patients with EF <40% are at a much greater risk not only of cardiac failure but of arrhythmic death. Isolated VEBs or short runs of VT are often innocuous and may be symptomless. Although highly frequent VEBs may have a poor prognosis, this is because they reflect serious myocardial disease, not because they initiate arrhythmic death. Thus, in an attempt to identify the true risk factors, attention has been directed much more to the desynchronization of ventricular depolarization as indicated by the technique of signal averaging of successive electrocardiographic (ECG) records to detect continuing electrical activity in the wake of the QRS in what should be an isoelectric interval. Similarly, heterogeneity of repolarization can be detected by software measuring dispersion of QT intervals between different ECG leads. A more sophisticated investigation combining late potentials and spectral turbulence analysis to predict arrhythmic events after MI on 778 patients scored a sensitivity of 78%, specificity of 63%, positive predictive value 8%, and negative predictive value 98% (10). Identifying a patient at risk of arrhythmia, however, does not necessarily involve an obligation to treat, and the authors themselves concluded that "drug therapy and referral for revascularization procedures during the postinfarction period was at the discretion of the patient's physician."

PATIENT INVOLVEMENT IN DECISIONS

Since both inactivity and treatment involve risks that are difficult to quantify, there is a case for involving patients, insofar as they are able to understand the issues, in decisions about possible therapies. If they themselves are helped to assess risks and benefits and concur in the ultimate choice, their own "will to improve" is enlisted, which may be beneficial, and they are prepared in advance for the possibility of failure. Biologic variation ensures that every individual is to some extent unique, both in the manifestation of disease and in response to treatment, so that there can never be an ideal and universal procedure applicable to all. A degree of consensus, however, has emerged over the years about general principles for management of therapy in a number of areas, continuously adjustable as new drugs are introduced and interventions developed.

CARDIOLOGY IN PERSPECTIVE

Advances occur so rapidly that any publication can be no more than a snapshot of today, out of date tomorrow. Nevertheless, it is sensible to stand back and take a long view from time to time, assessing the historical development of the subject, to extract lessons for the future and jettison discredited hypotheses. For example, an ionic current regarded as fundamental to pacemaking 20 years ago was later found not to exist, and treatments once widely used have been abandoned.

Although scientifically based clinical cardiology is less than a century old, there have been landmark observations of relevance from much earlier dates. A book recently published by E. J. Wormer (11), unfortunately not yet available in translation, assembled a series of monographs on the lives and work of 36 pioneers in studies of the heart from 1761 (Valsalva) to 1975. It contained a foreword in English that is quoted here as it seems particularly relevant:

"The author of this volume is to be congratulated on the originality of his idea, to take a theme, the development of Cardiology, and to illustrate it with accounts of the work and lives of some of those who have contributed to it. Dr. Wormer himself comments that "one main theme is the idea of medicine as an international peaceful language that makes communications possible between people beyond any racial, national, or political borderlines." In this respect Art is no different from Science; appreciation of a painting or piano concerto transcends frontiers worldwide. In another sense, however, science exerts a fascination not shared with the arts, in that the truth, like Mount Everest, is *there*.

The heart actually does work in certain ways, and it is up to the investigator to discover what they are. A major problem is that the same experimental result can be interpreted in many different ways, but the author of an idea, if imbued with the scientific spirit, should be the first to congratulate a colleague who provides a more probable explanation of the data, even if his own theory is thereby demolished. The early alchemists, although some were motivated by greed and all believed much that is now unbelievable, formed an international brotherhood of devotees who shared their secrets and supported each other in the face of persecution.

Their co-operation led to an accumulation of empirical knowledge from which the science of chemistry emerged, even though the philosopher's stone eluded them. Long may such co-operation continue in cardiology. The perfect antiarrhythmic drug will not be found, but the combined efforts of chemists, pharmacologists and physicians have led to the introduction of compounds which have improved the quality and extended the duration of thousands of lives, to achieve which is, after all, the primary aim of medical science."

REFERENCES

1. Rosen MR, Breidhardt G. Foreword. *Eur Heart J* 1995; 16(suppl G):1.
2. Breidhardt G, Borggrefe M, Fetsch T, Böcker D, Mäkijärvi M, Reinhardt L. Prognosis and risk stratification after myocardial infarction. *Eur Heart J* 1995;16(suppl G):10–19.
3. Tavazzi L, Mortara A. Exercise training and the autonomic nervous system in chronic heart failure. *Eur Heart J* 1995;16:1308–1310.
4. Östman-Smith I. Reduction by oral propranolol treatment of left ventricular hypertrophy secondary to pressure-overload in the rat. *Br J Pharmacol* 1995;116: 2703–2709.
5. Borer Symposium. Panel discussion. *Am J Ther* 1995;2: 347–358.
6. Cowley AJ. The clinical impact of coronary artery disease: Are subjective measures of health status more relevant than laboratory-assessed exercise tolerance? *Eur Heart J* 1995;16:1461–1462.
7. Ilson BE. Cardiovascular monitoring in normal healthy adults. *Am J Ther* 1995;2:893–899.
8. Campbell RWF, Murray A, Julian DG. Ventricular arrhythmias in the first 12 hours of acute myocardial infarction. *Br Heart J* 1981;46:351–357.
9. Campbell RWF. Ventricular ectopic beats and non-sustained ventricular tachycardia. *Lancet* 1993;341: 1454–1458.
10. Mäkijärvi M, Fetsch T, Reinhardt L, Martinez-Rubio A, Shenasa M. Comparison and combination of late potentials and spectral turbulence analysis to predict arrhythmic events after myocardial infarction in the Post-Infarction Late Potential (PILP) study. *Eur Heart J* 1995;16:651–659.
11. Wormer EJ. *Syndrome der Kardiologie und Ihre Sch pfer.* Münch: Medikon Verlag, 1989.

Subject Index

Subject Index